COLS
WRIGHT STATE UNIVERSITY
UNIVERSITY LIBRARY
0 00 13 0387548 6

P9-AOU-895

After the Second Oil Crisis

After the Second Oil Crisis

Energy Policies in Europe, America, and Japan

Edited by
Wilfrid L. Kohl
School of Advanced International Studies,
The Johns Hopkins University

LexingtonBooks
D.C. Heath and Company
Lexington, Massachusetts
Toronto

Library of Congress Cataloging in Publication Data

Main entry under title:
After the second oil crisis.

1. Energy policy. I. Kohl, Wilfrid L.
HD9502.A2A33 333.79 81-47023
ISBN 0-669-04547-0

Published simultaneously in Canada

Printed in the United States of America

International Standard Book Number: 0-669-04547-0

Library of Congress Catalog Card Number: 81-47023

For my colleagues on the faculty and staff
at The Johns Hopkins Bologna Center, 1976-1980.

Contents

Preface and Acknowledgments

This project began as a series of lectures at The Johns Hopkins Bologna Center during 1979-1980, when I was director of the Center and its Research Institute. Following my return to the Washington faculty of the School of Advanced International Studies (SAIS), the project was expanded to include other contributions during 1980-1981. All of the earlier contributions have been revised to take account of subsequent events.

For support I am grateful to the Research Institute of the Bologna Center, and to The Johns Hopkins Foreign Policy Institute in Washington, D.C. As it turned out, this book has become a joint publication of both institutes. Through these institutes, the nationalities of the various contributors, as well as the countries treated themselves, this was a genuine European-American undertaking. However, no serious study could ignore Japan, a major actor in the international oil market, and a nation today considered by many a part of the Western advanced nations' economic order.

Appreciation is expressed to two recent SAIS M.A. graduates, Alan Asay and Gary L. Sojka, who provided research and editorial assistance and helped in countless other ways. Well before the end of the project they had become valued colleagues.

After the Second Oil Crisis

1 Introduction: The Second Oil Crisis and the Western Energy Problem

Wilfrid L. Kohl

The cut-off of Iranian oil supplies following the collapse of the Shah's regime in the autumn of 1978 unleashed a series of events in the world oil market in 1979 that are now referred to as the second oil crisis. Though different in many respects from the Arab oil embargo of 1973-1974, the second crisis was also a watershed in terms of its effects on international energy relations. The impact of the second crisis, and the directions of Western energy policies after the crisis are the subject of this book.

In contrast with the embargo in the first oil crisis, the second crisis involved a shorter period of scarcity. Beginning in December 1978, Iranian oil exports ceased for about sixty-nine days. Then production resumed in Iran at a somewhat lower level. However, increased production by Saudi Arabia and other producers soon made up for most of the shortfall, which was limited to from one to three million barrels/day (mbd) over a short period of time. Much more serious were perceptions by importing countries and their oil companies of future oil scarcity, which led to panic purchases on the spot market in order to increase oil stocks. Escalating spot-market prices beginning in early 1979 then drove up Organization of Petroleum Exporting Countries (OPEC) oil prices in successive increments.

Thus, the second oil shock was essentially a price crisis. The official price of Arab marker crude rose from $12.70 in December 1978 to $24 in December 1979, an increase of almost 100 percent. Other OPEC member prices varied above that level, and spot prices soared above $30 a barrel. The rapid rise in oil prices over such a short period of time placed a heavy burden on Western economies, further fueling inflation, slowing economic growth, and straining payments balances.

The crisis accelerated trends of structural change in the oil market, as discussed further in a later chapter. Tendencies on the part of OPEC producers to conserve their resources were reinforced, since higher prices allowed them to produce less and still earn the revenues they need. Direct OPEC sales to governments and government-sponsored oil companies increased by some 50 percent in 1979, and several European countries and Japan were heavily involved. The role of the international majors was thereby constrained, reducing their ability to sell oil to third-party companies. A major impact of the crisis was therefore felt by the independent oil companies. Moreover, the OPEC price structure became more complex. Many OPEC members set their own prices according to types of contracts or deals above the agreed benchmark price. All of these structural changes further reduced the efficiency of the world oil market and its ability to reallocate oil in future crises.

The collective Western response to the second oil crisis was weak. The International Energy Agency (IEA) had been primarily designed to deal with supply disruptions, not price crises. Thus, it had to improvise. At a meeting of its governing board on March 1, 1979, agreement was reached to reduce oil demand on world markets by 5 percent. But the agreement fell apart. Germany, Sweden, France, and other countries sent missions to the Arab states, and there was extensive buying on the spot market. The Tokyo Summit in June 1979 set modest import ceilings and group-oil-consumption targets for 1985. But the summit was poorly prepared. Neither the summit nor the IEA had devised a mechanism to monitor compliance. It proved difficult to translate group targets into national targets, reflecting the reluctance of countries with different energy endowments and domestic pressures to make tough decisions to reduce oil demand. Oil prices on the spot market continued to spiral upward. The European Communities (EC) which had also set import ceilings at Strasbourg, did not fare any better. A French proposal to control spot-market prices was rejected by Germany, the United Kingdom, and the Netherlands, countries which resisted intervention in the marketplace.

What developments have affected the international oil market since 1979, and what is the nature of the Western energy problem in the early 1980s? In 1979, oil made up about 55 percent of total energy consumption in the European Communities; the comparable figures were about 75 percent for Japan and 26 percent for the United States. Western Europe imported about 84 percent of the oil it consumed in 1979 (compared to 99 percent in 1973), Japan continued to import 99 percent of its oil, and the United States imported about 44 percent of its oil (compared with 35 percent in October 1973). The bulk of the oil imported to Western Europe (about 8.7 million barrels/day), and to Japan (about 3.4 mbd), and about one-third of American imports (about 3.1 mbd), came from the Arab members of OPEC in the Middle East/Persian Gulf area, one of the world's politically least stable regions.

Although it produced a mini-shock, the Iran-Iraq war, which broke out in the summer of 1980, did not cause a third oil crisis for a number of reasons. This time the disruption occurred in a situation of surplus on the world oil market, for the price effects of the 1979 crisis had lowered Western oil demand. Saudi Arabia again increased production to make up part of the shortfall. The aggregate supply loss for IEA countries was estimated at less than 2 mbd. Moreover, most industrial-country oil stocks were high. This time the collective Western response was also more effective. Beginning in October 1980 the International Energy Agency gained the assent of member governments to take measures to discourage spot-market purchases and draw down collective oil stocks. Further measures were taken in December to reduce oil demand, thereby heading off an oil-price spiral. The economic recession and other factors moved Germany, Japan, and other countries to support the agreement. The IEA also played a useful informal role in assisting two countries that had suffered heavy supply losses, Turkey and Portugal, to gain access to oil supplies. By January 1981 some oil exports had started again from Iraq and Iran, although the war continued in a low key.

In 1980 the Organization for Economic Cooperation and Development (OECD) countries, assisted by persisting low levels of economic growth and energy conservation, were able collectively to reduce their demand for oil by 7 to 8 percent. The downward trend in Western oil demand continued during 1981, contributing to an oil surplus that was predicted to last into 1982, perhaps longer. Reinforced by advances in energy substitution, the oversupply in the international oil market was also rooted in increased non-OPEC oil production in Mexico and the North Sea, and Saudi Arabia's decision to produce oil at above-normal rates in order to force its price policy on other OPEC members.

The decline in OPEC's share of the world oil production and the oil glut appeared to increase Western leverage over OPEC andled to a chaotic OPEC meeting in Geneva at the end of May 1981, when most members decided to freeze prices at then-current levels ranging from $36 to $41 per barrel, and to cut production by at least 10 percent. Saudi Arabia, on the other hand, holding fast to its strategy of trying to lead the cartel toward a unified price structure, but failing to rally support, decided to maintain its price for benchmark crude at $32 a barrel and to continue to produce 10 million barrels of oil per day. The Saudis were obviously concerned that oil prices were rising too rapidly and encouraging production of alternative fuels, a trend that, if continued, could eventually price OPEC out of the market.

Meeting again at the end of October 1981, OPEC finally agreed to restore a unified price structure that it had not had since 1978. The Saudi strategy had paid off. In return for raising its price for benchmark crude to $34 a barrel, Saudi Arabia had compelled other members to reduce their marker prices to the same level. The new benchmark price was supposed to last through 1982. Premiums for higher-grade crudes would be permitted up to $38 a barrel, although the premium structure was to be reviewed again at the end of the year. Saudi Arabia also agreed to cut its oil production by 1 mbd, to 8.5 mbd. The open question was whether the other OPEC members would be able to regain their lost market shares in the face of a continuing slump in world demand, thereby laying the basis for new oil-market stability. The Saudis might be expected to reduce production still further if necessary to overcome the oil glut and hold the new benchmark price.

The oil-market surplus of 1981 clouded a situation that continued to be fragile. It could lull the industrial oil-consuming countries into believing that continuing efforts at conservation and developing alternative energy sources were unnecessary. A gradual economic upturn might tempt Western nations to increase oil imports again, which would increase their vulnerability. Although some progress has been made in reducing Western dependence on OPEC oil, much remains to be done at both the national and international levels to manage the energy transition and achieve the structural changes necessary to significantly reduce our reliance on oil over the longer run.

Energy dependence will continue to pose a security problem for Europe, the United States, and Japan in the 1980s and 1990s. These countries will still rely heavily on imported oil. War or revolution in key OPEC countries could produce a new crisis. The ultimate nightmare would be a revolution in Saudi Arabia. There is also a danger of Soviet intervention in the Persian Gulf. Or Saudi Arabia, which has enormous clout by virtue of its giant reserves and ability to produce 40 percent or more of OPEC's production, could simply decide to produce less oil, which would immediately have a large impact on world supply and drive oil prices up again.

Looking ahead, Western nations face the problem of improving their capability to cope with future oil crises (which will very likely occur)—internationally, through stronger institutional arrangements at the intergovernmental level; domestically, by building up emergency oil stockpiles and devising new policies for their management. Moreover, more effective policies need to be formulated at the national level to encourage switching to other fuels (mainly coal, gas, and nuclear energy), and energy saving in order to reduce further oil imports. Though free market pricing is a vital element in promoting the energy transition, it seems clear that governments will also have to play a role. International organizations, such as the IEA and the EC, can also assist by providing a forum to coordinate policies and prod governments into action. Over the long term, it is important that new energy technologies be explored and developed. Those that prove feasible will provide future options when they come into commercial production, probably in the next century, to diminish as much as possible Western oil dependence.

The chapters presented here examine from various angles the impact of the 1979 crisis on the Western energy system, especially its consequences for important industrial countries, the oil industry and other energy sectors. An OPEC view offers a critical perspective on the crisis. But the emphasis is on energy problems in the industrial world and future options for Western energy policies in the early 1980s. All of the authors are recognized experts and a number of them are participant-observers who have worked in governments or international organizations in the energy field.

The chapters in Part I focus on specific energy sectors. The first two deal with changes in the international oil market. In explaining OPEC's view, Fadhil Al-Chalabi, OPEC's Deputy Secretary General, points out that the second crisis was not so much a crisis as a continuation of the process that began in 1973-1974, correcting an imbalance between long-term supply and demand for oil, and optimizing utilization of a finite resource. The second crisis, which was caused by market disruptions and not by OPEC, demonstrated that OPEC's post-1973 pricing policies were not sufficient to achieve the required balance in the energy market. In his view, both producers and consumers will benefit in the long run from a more realistic price of oil, one more comparable to that of alternative energy sources.

Ian Torrens reviews trends in oil supply and demand in Europe in the 1970s and changes in the structure of the international oil industry. His chapter provides an insider's view of the role of the International Energy Agency, which was established after the first oil crisis to reduce the vulnerability of the consuming industrial countries to future oil disruptions by an emergency oil-sharing system, and to promote the restructuring of Western energy economies away from primary dependence on oil.

Turning to other energy sectors, the present state and future prospects of nuclear power and its role in meeting energy needs are explored by Georges Delcoigne. Problems of nuclear trade and the link between nuclear power and nonproliferation are analyzed, with particular attention given to the implications of the Carter Administration's policies as seen from a European perspective.

The most feasible short-run alternative to oil is coal, according to William F. Martin, who explains how coal production and coal trade could be expanded to meet the objective of doubling world coal use by 1990 and tripling it by the end of the century. In reporting on a recent systems study conducted for the IEA, Neils de Terra discusses current efforts to develop an industrial-country group strategy for energy research and development and examines which technologies offer the most promise between now and the year 2000 and beyond.

Part II treats the energy policies of seven industrial countries after the second oil crisis. It also includes chapters on the efforts to form a common energy policy in the European communities, and on the politics of participation in energy-policy formulation. In broad terms, the chapters in this part seek to illuminate the energy situations of the countries concerned, their structures, the impact of the second oil crisis on their energy policies, and problems of their policymaking processes.

Among the countries considered, it should be pointed out that three are major energy-producers—the United States, Canada, and Great Britain. The United States, of course, is a special case, since it consumes more energy per capita than any other country in the world. It is also the home of several major international oil companies. Yet the United States is the world's largest oil importer in absolute terms, and has had great difficulty until recently in limiting oil imports. The other countries, the European middle powers and Japan, are dependent on external energy resources to a considerable degree.

Several chapters illustrate that basic features of political cultures and political systems can be critical elements in determining whether a country is able to cope effectively with its energy situation. For example, in the case of the United States, as Paul S. Basile points out, fragmentation of power in the political system prevented passage of important energy measures proposed by the President and produced a domestic-energy stalemate. The second oil crisis provided the impetus needed to break that stalemate in the third and fourth years of the Carter Administration. The Reagan Administration is further reforming U.S. energy policies.

In Canada, as discussed by John F. Helliwell, the second oil crisis had the opposite effect, undermining a fragile system of price negotiation and compromise that had existed between the resource-rich western provinces, primarily Alberta, and the federal government. The energy issue, especially how to distribute large windfall profits, intensified an impending constitutional crisis in Canada's federal system, throwing energy policy into a stalemate.

In Great Britain, a resource-rich country since the opening of the North Sea oil fields, the question of how to distribute windfall gains accruing to the national oil and gas companies and the government has been tempered by the postwar British tradition emphasizing the provision of state services. The Thatcher government would like to use these revenues to restructure British industry, but may find it difficult to do so, as Nigel Lucas argues.

Economic factors, including the structure and health of the overall economy, the nature of economic policy choices, and leadership have been especially important in explaining the impact of and responses to the second oil crisis in countries that are heavily resource-dependent.

In France, the higher oil prices created a major deficit in the balance of payments and spurred a recession. However, the government of President Valery Giscard d'Estaing decided in 1980 to sharply reduce the country's oil consumption over the next decade, to diversify energy sources, and to accelerate energy saving—worthy objectives, argues Guy de Carmoy. The centralized French economy with major public enterprises in the energy sector will ease the government's task in strengthening French energy policy, which may be slightly reoriented following the election of a socialist government in 1981.

In Germany, the world's third-largest oil importer, the oil bill had by 1979-1980 yielded the first balance-of-payments deficit in over a decade and contributed importantly to rising inflation. Dieter Schmitt describes Germany's free-market approach while emphasizing the need to reduce dependence on imported oil by expanding the use of coal, and, if domestic opposition can be overcome, nuclear power.

Italy, too, suffered a serious trade deficit and accelerated inflation following the oil price hikes in 1979. A country dependent on imports for 98.7 percent of its oil, Italy must diversify its energy sources and develop nuclear power, renewable resources, and expand its use of coal, contends Umberto Colombo. But frequent government crises and erratic parliamentary consideration have made it difficult for the country to formulate sound energy policy.

Despite its position as the world's second largest oil importer, Japan has weathered the second oil crisis more successfully than other nations, explains Ronald A. Morse. The key seems to have been the general strength of its economy and its resistance to price controls. Grounded on close cooperation between government and industry, Japan is pursuing various options to mitigate its weak resource position, including conservation; the expanded use of liquefied natural gas, coal, and nuclear power; and development of new energy technologies. However, Japan is weaker in emergency preparedness and seems less willing than it perhaps should be to participate in coordinated international energy strategies.

The 1979 crisis gave some impetus to efforts of the European Communities to shape common energy policies, especially in the field of energy saving. But, on the whole, the EC has not managed to provide much more than targets for reducing oil consumption and fuel switching and a framework against which to coordinate national energy policies. The Communities have been hampered by institutional weakness and lack of a sufficient mandate. Different resource endowments of the member countries and the political sensitivity of the energy issue for national sovereignty have also impeded an integrated approach.

In the concluding chapter, Volkmar Lauber surveys new forms of political participation and protest that have evolved in a variety of Western countries in response to energy issues. He discusses citizen groups that have been formed outside normal political channels to protest decisions of the state, as well as forms of participation offered by states. An important dilemma is posed: In most democracies existing political structures and organized interests tend to favor continued economic growth, rising living standards, and hence greater energy use. If energy conservation is to be taken seriously, existing structures will have to be challenged.

Part I
Developments in Key Energy Sectors

2 A Second Oil Crisis? A Producer's View of the Oil Developments of 1979

Fadhil Al-Chalabi

The oil developments of 1979 are often referred to in the developed oil-importing countries as the *second oil crisis,* just as the events of 1973-1974 are commonly accepted there as the *first oil crisis!* This consumers' approach is conceptually different from that of the producers when it comes to explaining the great structural changes of the 1970s in the international oil industry, and more generally in world energy relations.

With the important and rather abrupt increases in the prices of OPEC oil, consumers found themselves suddenly paying a real price for their oil imports, after decades of having been used to unusually high growth rates based on an almost costless form of imported energy. Naturally, such a sudden change toward a substantially different cost structure tends to create, in the short term, a disruption in trading and payments systems, before consumers adapt their economies to the new structural change in the relationship between energy and economic growth. More important for them, such price increases seem to create an adverse transfer of resources that affect economic growth and welfare.

For producers, on the other hand, these developments were no more than a continuous process of correcting imbalances related to economic and social development, and the optimum utilization of a finite resource. These developments meant for producers the restoring of inalienable rights of sovereignty over their natural resources, from which they had virtually been deprived during the decades of domination by the multinational oil companies. Such developments also meant a radical political change to allow producers to optimize the utilization of their oil resources toward achieving a higher pace of economic and social development. Finally, direct control over their finite resources places producers now in a more favorable position to achieve a better balance between their present requirements and those of future generations.

Therefore, what may be called in the West the *first energy crisis* was for the producers a *revolution,* because the events that led to the quadrupling of oil prices during that period heralded a new era of self-rule in the manage-

The views expressed in this paper are those of the author and do not necessarily represent official positions of OPEC.

ment and pricing systems of raw materials. It replaced the old relationship of domination of raw-material-exporting countries by the consuming countries.

Until the beginning of the 1970s, the setting of the so-called posted prices (that served as a basis for calculating the per-barrel government share) was the exclusive prerogative of the multinational oil companies through the concession system inherited from the days of colonial rule. It is now admitted, even in the West, that prior to the 1970s, those prices were set at such an artificially low level that the real cost of acquiring imported oil in Europe and Japan was negligible. The per-barrel government take was even smaller as it was only marginally more than half of the posted prices. In spite of various amendments to the fiscal regime of the oil concession, in agreement with the companies, the government take did not increase substantially from $1. Even with the improvements that were brought by the Teheran and related agreements of 1971 and 1972, the government take did not surpass $2 per barrel until October 1973.[1] For a community so heavily dependent on a single and finite resource, this situation had more than one adverse effect.

First, the extremely low cost of acquiring a barrel of oil from the OPEC area led, during the postwar period, to a fast and unprecedented shift in the energy structures of the developed oil-importing countries of Western Europe and Japan from coal to oil, with the consumption of imported oil from the OPEC area doubling almost every 6 years. A corollary to that was an exceptionally high depletion rate of the oil resources in the producing areas, as measured by the dramatic decline of the oil-resources life span. Whereas , in the mid-1950s there were enough recoverable reserves in the Middle East to sustain about 130 years of production, the life span of those reserves dropped to 70 years in the mid-1960s and to about 40 years in the mid-1970s (see figure 2-1). This happened during the years of the great oil discoveries in the Middle East and Africa (the 1950s and 1960s) when proven and recoverable reserves in the OPEC area more than tripled! Replacing depleted resources involves high financial and technical risks, and implies prohibitive costs as well as extremely long lead times, which cannot be sustained by the producing countries, still at a stage of early economic and social development.

Second, the meager financial resources generated from the old system represented a real constraint on development of the OPEC countries, because what they were getting from an enormous international trade of their oil was not even enough to meet their day-to-day expenditure requirements. Governments of the consuming importing countries, together with oil-trading companies, received the bulk of the economic rent generated from oil, leaving the owners a meager share.[2]

Third, cheap oil in the past resulted in great world energy imbalances. Whereas oil and gas account at present for about 70 percent of the world's

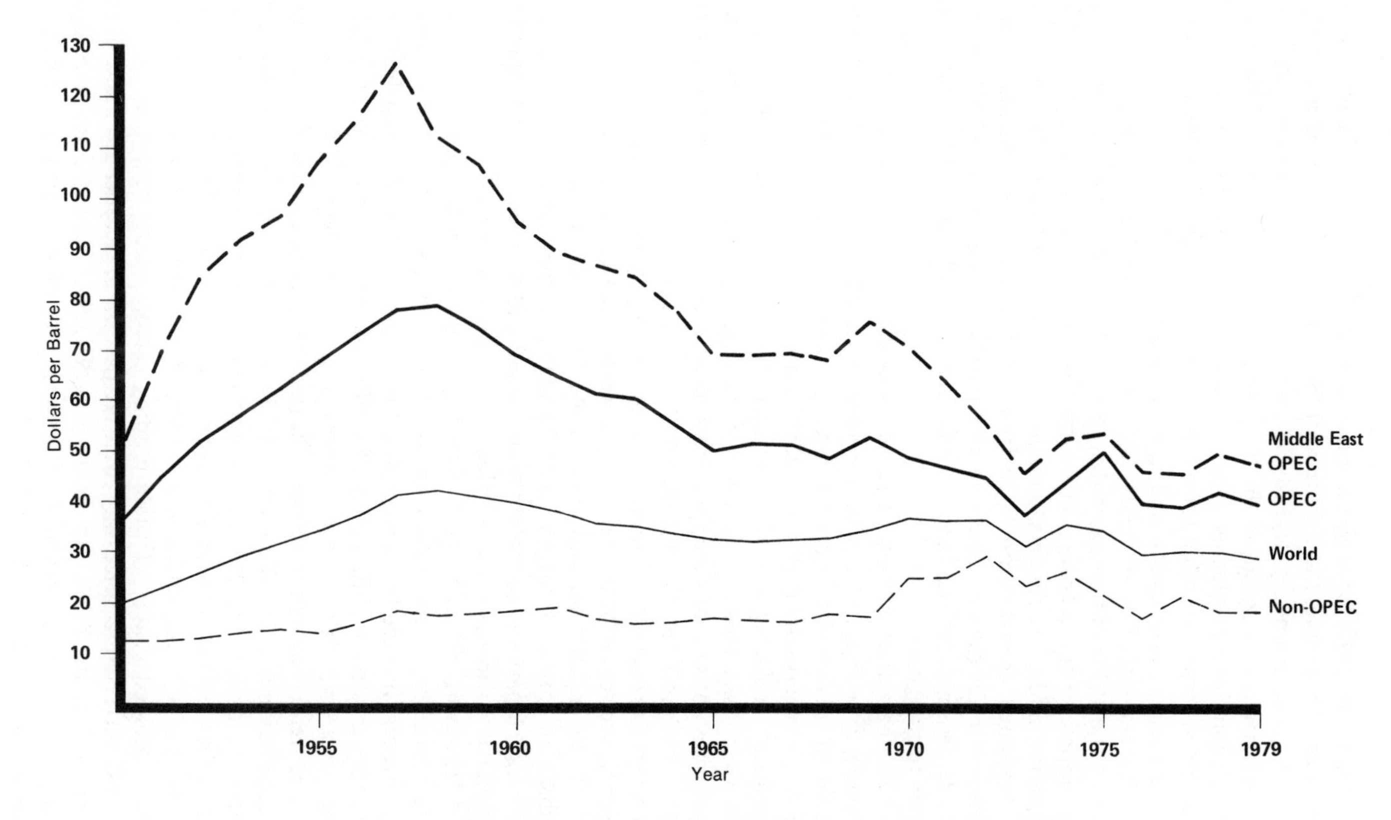

Figure 2-1. Oil-Reserve to Production Ratios, 1950-1979

total requirements, they represent not more than 20 percent of global energy resources. More than half of the world's oil requirements are met by OPEC exports. If such a heavy dependence on OPEC oil for the world's economic growth persists, it could, in the long run, add to the growing pressures on OPEC members' producing capacities, and eventually reduce their ability to meet future world demand for oil, including their increasing demand for oil for domestic consumption and industrialization.

The reversal of those trends was therefore inevitable and OPEC's actions during the 1970s did no more than correct basic imbalances of energy and development: cheap energy meant speeding up growth in the oil-consuming countries and arresting development in the oil-producing countries through a continued deterioration in their terms of trade. It meant also greater dependence by the consumers on imported energy and hence a faster pace in the depletion of an irreplaceable finite resource needed by the producers for their future generations.

The partial correction of those imbalances not only benefited the oil-producing countries, but also the consuming countries. One major positive result of this corrective process was to create, through the increase in the prices, an energy-saving consciousness for the first time in the industrialized countries, replacing attitudes that encouraged the wasting of energy in the era when it was cheap. This is how, for example, Western European oil consumption has virtually stopped growing since 1973. In contrast, in the five years that preceded the "revolution," the average annual rate of increase of Western European consumption was around 7 percent. These countries also became conscious of the necessity to create a new energy structure, through which dependence on imported oil would be reduced and investments in substitute forms of energy could be made. Some measures have already been taken partially to replace oil with coal and programs were adopted for greater investments in non-OPEC oil and nuclear energy.

However, OPEC's post-1973 price policies meant that the initial "revolutionary" momentum was not kept going, as it should have been in any process intended to correct structural imbalances. One major feature of the conservative pricing policy of OPEC was the adverse movement of its real price vis-à-vis the nominal prices. Between 1973 and 1979, world inflation increased by about 100 percent (using the Organization for Economic Cooperation and Development (OECD) export price indexes). In addition, the U.S. dollar, the unit of account denominating a barrel of oil, kept depreciating. Yet against this continued deterioration in the purchasing power of its barrel, OPEC did not adjust its price in dollar terms to offset the combined results of these eroding factors on its real price. Instead, OPEC corrected its price only twice (10 percent in October 1975 and 10 percent in January 1977). By January 1979 OPEC increased its price again marginally with the intention of achieving an average 10 percent price increase for the whole year of 1979.

As a result of this policy of declining oil prices in real terms (see figure 2-2), OPEC has made the imbalance appear again, because declining real prices meant a new deterioration in the terms of trade of OPEC member countries, and, above all, tended to slacken efforts in the western countries to save more energy, witnessed by the down-scaling of programs for investment in alternative sources of energy and the sharp increase of oil imports in the United States (which was recently importing almost half of its oil consumption). Figure 2-2 clearly indicates that it is only with the price increases during 1979 and the first half of 1980 (and that is what is essentially called the "second oil crisis") that the price of oil became equal in real terms to that of January 1974. In other words, if prices in dollar terms were continually raised since January 1974 to offset the eroding effects of world inflation and the dollar depreciation in order just to keep intact the real purchasing power of the barrel in 1973 dollars, that is, without adding any real increases in the price, oil prices would have automatically reached the mid-1980 levels.[3]

By administering its prices at lower levels in real terms, OPEC believed it was helping the world economy attain higher growth rates. But, as the events of 1979 showed, the cost of that help was very high in terms of energy balances; for as soon as the Iranian oil crisis erupted, the OPEC price decision of January was completely shattered. More important, subsequent developments during the year proved that OPEC's price administration was ineffective when the market was boiling, and stricken by a great

Figure 2-2. The Evolution of Nominal and Real Prices of Oil since 1973

sense of panic and insecurity. It was the market, and not OPEC, that was actually taking over the process of pricing oil. What OPEC was trying to achieve in the course of 1979 was to put some brakes on a price race, in which higher official prices set by a member country were soon outdated by increasing bids in the market.

Assuming that these developments were temporary, many of the oil producers began adding to their OPEC official prices surcharges and premiums that were, presumably, to be removed once the market regained its stability. Price decisions of 1979, however, had little effect in restoring stability, and OPEC conference meetings during that year did no more than add to official OPEC prices only a part of the market increase. As a result the gap between OPEC official prices and market prices was kept wide. Spot-market quotations for crude oil often reached more than twice OPEC official prices. Similarly, in free markets such as Rotterdam, prices of refined products, if netted back into FOB (free on board) values for crude exported from the main producing areas, revealed exactly the same gaps.

In an effort to stabilize the market, OPEC decided in its June 1979 session in Geneva to put a ceiling on its price structure, so that no crude in any member country would be selling beyond the ceiling price. Surviving for only one quarter, that ceiling was again shattered by the panic in the market. The result was that instead of a price of $14.55 per barrel that the marker crude (Arabian Light) would have reached in December 1979, had the formula of 10 percent price increase decided by OPEC for the whole year been implemented without market intervention, the official price of Arabian Light reached $24 per barrel and then to $28 during the first half of 1980.

The price confusion was not confined to the growing difference between OPEC prices and market prices, but was also reflected in the increasing and heterogenous gaps between the OPEC prices themselves (the OPEC price differentials). In spite of its many imperfections, OPEC's price structure before these developments reflected a reasonable degree of homogeneity. Thus, the light, low-sulfur Saharan Blend, for example, was priced in January 1979 only $1.50 per barrel higher than the marker crude, whereas the heavier crudes were priced at lower levels. The market convulsions had their own effects on these differentials. By the middle of 1979, the differential of Saharan Blend had shot up to $8 per barrel or more, whereas the heavier crudes were priced at higher levels than the marker crude.

The most important new development in the inter-OPEC price relationships during 1979 was the independent line taken in the pricing of Arabian Light as the OPEC marker crude, which constituted a real departure from the traditional process of price making within OPEC since the days when it took over the fixing of oil prices. The pricing of OPEC's marker crude was traditionally based on a unanimous decision by the OPEC Conference, with its producer, the Kingdom of Saudi Arabia, playing a major, if not a crucial role. Under the new system, the price of Arabian Light came to be unilaterally decided by Saudi Arabia. This new trend in price making, if it persists,[4] shall have a far-reaching impact on the OPEC price structure, as it can allow for wider price gaps between the marker crude and the other OPEC crudes, especially in times of market tightness. It could also enormously affect the process of price making within OPEC, particularly in a time of market glut.

But what happened to justify such market disruptions, which created a price situation that could not be controlled by OPEC? Certainly it was not because of a short-term imbalance between world supply of OPEC and non-OPEC oil and consumption. In spite of the sudden dissapearance of over 5 million barrels/day (which was the Iranian exported production), the market did not in fact actually suffer from any physical shortage. The Iranian shutdown was counterbalanced by increases in production by many important OPEC producers, especially Saudi Arabia, Iraq, and Kuwait. Total OPEC production in 1979 was in fact higher than in 1978 by about 4 percent in the face of stable, if not declining, oil consumption levels in the industrialized countries. In other words, the increased demand for OPEC oil was greater than warranted by actual consumption figures. This was especially so if we take non-OPEC production, which rose by more than 5 percent during that year.

The unusual phenomenon of these imbalances can be attributed to some factors of a short-term nature. Most important of these is the abnormally high oil inventories kept by the industrialized countries. By January 1980 the total stock buildup outside the centrally planned economies (CPEs) reached about 5.3 billion barrels. This unprecedented high level of stock accumulation was equivalent to almost half of total OPEC production during 1979. It is more than the entire recoverable reserves from a small producing country such as Qatar. In 1979 alone these stocks increased by more than one billion barrels, or three times as much as the annual production of Algeria. What helped operators partially in accumulating this increase in stocks was the ever-appreciating value of those stocks compared with their original cost. Strengthening the buyer's position in the market was a greater incentive for such hectic stock replenishment, and their feeling of supply insecurity because of political instability.

The market tension of 1979 can also partially be explained by the changes that took place in the nature of the international trading systems and contractual relationships. The major oil companies, as is known, used to supply third-party buyers with substantial quantities of the crude oil at their disposal, over and above their own refinery and distribution-network requirements. The developments in the government-company relationship, which resulted in governments taking over the management of their oil industries, either through nationalization or other arrangements, reduced the quantities of crude available to the major oil companies substantially in favor of OPEC national oil companies. With the Iranian termination of previous long-term lifting arrangements with the ex-consortium shareholders, as well as nationalization in Nigeria, the major oil companies had to cut off their contractual crude supplies to third-party buyers, mostly independent refiners. As a result, in order to meet their commitments, these independent refiners were willing to buy at any price in the market. This happened at a time when the spot market had grown substantially from the increased entry of many operators and traders wishing to profit from market fluctuations. One result was that many speculatory transactions took place on cargoes that never actually changed hands.

A further more immediate factor that helped to aggravate market instability was the change in the nature of demand for both oil products and incremental crude-oil supplies. Following the great structural developments in the oil industry during the 1970s, demand for heavy products dropped more sharply than for lighter products, as the conservation measures in the consuming countries, as well as the fuel-substitution process, hit more at the heavy ends of the barrel than at its lighter ends. The light products, such as gasoline and gas oil, were not as affected by the post-1973 oil-price explosion as was fuel oil. Against this structural change on the demand side, there was an inverse change on the supply side, since the incremental quantities available from OPEC for trading tended to be the heavier crudes. Extra-light-oil production in OPEC is relatively small compared with that of heavy oil, and there were also other constraints on light- oil production, either because of conservation measures such as in Saudi Arabia (the famous 35/65 production ratio), or merely because of physical limitations. Market tensions were created by this development because refinery configurations were still unchanged from the past pattern based on high demand for heavy oil. Also the Iranian shutdown meant the disappearance of substantial quantities of light crude (half of its production) that added to the nervousness of the market.

Many other short-term aggravating factors can be added to the situation, but cannot alone explain this unusual market phenomenon. More important are the long-term factors that the developments of 1979 brought rather dramatically to the surface. The Iranian oil crisis made the world suddenly realize that OPEC's capacity to produce was not without limitation, as it was often thought to be in the past. Basically, therefore, the oil developments of 1979 can be explained by a feeling of insecurity over supply that governed a panic-stricken market, as if the world suddenly discovered the existence of a long-term imbalance between supply and demand for oil in international trade.

The market nervousness revealed a new fact, that most OPEC countries were already producing at capacity and were unable to surpass this limit.

Going beyond capacity implied huge investments and long lead times in the case of most OPEC countries, such as Algeria, Nigeria, Libya, Venezuela, and so on. Only a few, such as Saudi Arabia, Iraq, and Kuwait, were able to take the incremental demand, but not, however, without increasing technical constraints and additional costs. In other words, OPEC capacity cannot be increased easily and cheaply, with the result that output limitations in the OPEC area can pose a serious threat to the long-term balance between supply and demand.

In addition to the physical-capacity constraints, there exist other production constraints of an economic nature in almost all OPEC countries that will tend to reduce available OPEC oil in the future. The growing trend toward greater conservation of reserves for future generations is discouraging overproduction from existing reserves. On the other hand, producers find generally little incentive to invest in increasing production capacity (either through exploration to discover new oil or through enhanced recovery techniques). There is, in fact, a widespread conviction in OPEC countries favoring the slowing down of depletion rates in reserves and preserving oil for the future development of their societies. The declining ratios between reserves and production, would, if no new oil is added, further reduce the life span of reserves in countries that have already reached capacity limits. In those countries able to increase production, the capacity to spend their incremental oil income on economic and social development is limited, as their absorptive capacity for development is constrained by many infrastructural and human bottlenecks. For them, therefore, producing more than is needed for development would mean accumulating surplus money that is increasingly vulnerable to the great risk of erosion in value caused by world inflation and currency fluctuation.

The general production philosophy gaining momentum nowadays in the OPEC area is to gear production levels to the optimum rates of economic and social development. Any production level beyond those rates would simply mean the accumulation of paper money that cannot be converted into real assets. In other words, for countries with limited prospects of speedy development of limited possibilities for structural diversification, producing more than is needed for development amounts to a sheer waste of national capital that will be badly needed in the future.[5]

In other oil-producing countries where the absorptive capacity to spend on development is higher, increasing oil production capacities would involve greater risks for their economic and social development. Apart from Iraq (where the capacity to spend more on development is high and where there is a huge potential for increasing oil production), most of the other countries with relatively higher levels of development needs suffer from physical limitations that prevent increases in capacity. This is why most of these countries are in deficit and have to borrow in the international capital market. In these countries it is extremely difficult to invest in increases in the oil production capacity, because the more these countries spend on finding new oil, the less money they can spend on their development programs, which could result in jeopardizing their long-term development prospects.

The underlying factors of an apparent second oil crisis in which oil prices in current dollars more than doubled in one year, were mainly related to the existence of an inherent imbalance between long-term supply and demand for OPEC oil. It is true that this imbalance was dramatically brought to the surface by some short-term aggravating factors, such as the unusual stock buildup by the oil lifters. But those latter factors were also an indication of a structural imbalance that required adjustment. This time, however, the adjustment was made by market disruptions rather than by OPEC. The fact that OPEC appeared almost unable to control the market could be taken to indicate that, while it succeeded in its initial price revolution to correct an old imbalance of wealth-sharing, it failed afterwards to sustain that initial correction process. Its post-1973 pricing policies were not adequate to achieve the required structural energy balances. Declining real prices had the effect of decelerating a process of structural change, which was initially triggered by the price explosions of 1973-1974.

The oil events of 1979 were no more, therefore, than a signal for these adjustments. Now that oil prices have risen in real terms beyond their levels of 1973, consumers will have more incentive to increase their efforts toward conservation, fuel substitution, investment in other sources of energy, and so on. It is obvious that only through higher prices in real terms can such endeavors be successfully achieved.

Like the first energy structural adjustments of 1973, what is called a second oil crisis is, therefore, again a process of correcting imbalances, this time created by the market and not by OPEC. The question, however, is whether it is in the interest of the international community and world economic growth, as well as the development of the oil-producing countries themselves, to leave the realization of adjustment to market forces, with all their dramatic and convulsive effects in the short term, or to resort to a planned approach whereby the price of oil would follow a certain predetermined path aimed at achieving a balance. Instead of waiting for the market to alleviate the pressure on OPEC capacity, OPEC can preempt such an eventual adjustment by adopting a long-term pricing policy based on gradual real increases in the price of oil in which supply/demand adjustments would be phased out in a stable and healthy market. In determining the size of its increases in real terms, OPEC could aim at equalizing, in the long term, its price with the cost of alternative sources of energy, particularly liquefied coal. Prices could first be adjusted periodically to offset the erosive effects of world inflation and currency variations, in other words, to prevent any erosion in the purchasing power of the barrel, and then by adding such gradual and small increases in real terms to the price within a planned time horizon in which oil prices in real terms would be on a par with the cost of those alternatives. Such declared pricing policies, which would assure a certain degree of predictability in price movement, would encourage viable investments in alternative sources of energy. In this way balances in the energy market could be achieved, pressure on OPEC production in the future be alleviated, and a more balanced energy structure in the world be realized.

Whether the events of 1979 are a second oil crisis as the consumers see it, or a new process of correcting imbalances, as the producers see it, both producers and consumers would in the end stand to benefit mutually from this process. It is only through higher real prices that world energy and development problems can be gradually solved. Cheap oil will serve no one's long-term objectives, as it was cheap oil that was behind the very high rates of the resource depletion in the 1960s that brought about the present long-term imbalance between supply and demand; just as higher prices were behind the conservation efforts that were made after 1973.

Notes

1. See details about posted price movements of the OPEC oil in Fadhil Al-Chalabi, *OPEC and the International Oil Industry: A Changing Structure,* Part Three (Oxford University Press, 1980).

2. The famous "OPEC Barrel" of the 1960s gives a good indication of the very unfair distribution of wealth generated from a barrel of crude between its users and owners. End-consumers in Western Europe used to pay $12 to $13 for a composite barrel of oil; of that the producer's share was less than $1 or about 7 percent. More than half of that price represented taxes levied by the governments of the consuming countries, while the remainder represented companies' profits and various values added generated in the different stages of the oil trade.

3. Those rather conservative results of the measurement of the relative movements in real terms of the OPEC price are based on the OECD Export Price Indexes. Many experts in the oil-producing countries, however, contest conceptually this measurement and believe that what OPEC is actually incurring as a cost for the imported inflation is much higher than what is suggested by those indexes. For this purpose an effort was made within OPEC to construct an index of its own to measure the imported world inflation (called the OPEC Import Price Index). According to this index the impact of the imported world inflation on the purchasing power of the barrel is more than twice that of the OECD price index.

4. Subsequent oil developments in 1980 indicate that the price heterogeneity resulting from the departure from the traditional process of OPEC price making, would persist only in times of market tightness. In June 1980, when the market still showed signs of tightness, OPEC's decision on pricing implied an actually $4 range for the price of marker crude, because of the existence of a price ceiling of $32 for market crude against its actual price of $28. By the middle of September when the market tightness had disappeared, OPEC succeeded in partially achieving price unification with a single price of $30 for marker crude. However, when faced with a new tightness in the market, the multitiered price structure reappeared in the Bali OPEC Conference price decision, with a $4 price range and a ceiling of $36 per barrel, versus an actual price of $32 for Arabian Light.

5. See Fadhil Al-Chalabi "The Concept of Conservation in the OPEC Member Countries," *OPEC Review,* 3:3 (Autumn 1979).

3 Oil Supply and Demand in Western Europe, the Oil Industry, and the Role of the IEA

Ian M. Torrens

As we begin the 1980s, the international price of crude oil is more than fifteen times what it was at the beginning of the seventies in current money, which means very approximately seven times in real terms. Given the importance of energy, and in particular of oil, as an economic input—indeed a necessity of life—this is a very fundamental change in relative values that implies significant readjustment of our pattern of consumption. And the change has not been smooth. It happened essentially in two discrete shocks—one at the end of 1973 and the other in 1979.

Significantly, despite this sevenfold increase over the decade, the real price of oil actually declined during some five out of the last ten years. This instability has made a rational response to the energy problem on the part of individuals and governments extremely difficult, so that as we enter a new decade, some adjustment has taken place to the higher relative price of energy, but not as much as might have been expected given the magnitude of the first price shock and the six years that have elapsed since it occurred.

Between 1975 and 1978 energy receded into the background of public awareness, as individuals discovered it was plentiful and getting relatively less expensive, and as governments had to deal with the more pressing problems of inflation, recession, and unemployment. As recently as the autumn of 1978 the conventional wisdom foresaw plentiful supplies of oil until the second half of the 1980s at the earliest. Some forecasters postponed any oil crisis indefinitely.

Then, with the closing days of 1978, the future arrived earlier than expected, precipitated by the revolution in Iran that caused an abrupt termination of its exports. The resulting oil crisis simmered on throughout 1979, and it cannot yet be said that its consequences are clear, or that the more than doubling of oil prices during 1979 brought an end to the upward price spiral. A stabilization of prices in 1980 seemed to be indicated by the global oil supply/demand balance until the onset of war between Iran and Iraq again upset the fragile oil-supply picture.

The opinions expressed in this paper are those of the author and do not necessarily represent the views of the OECD or of the governments of its Member Countries.

Add to this the implications of the Three-Mile Island accident in the United States for nuclear power—the only indigenous energy resource that some countries can develop on a significant scale—and it is hardly surprising that energy remains high on the international political and economic agenda.

The main purpose of this chapter is to set Western Europe's energy situation in the global perspective, and to describe the role of two quite disparate actors on the energy stage, namely the oil industry and the International Energy Agency (IEA). To create this perspective requires first a brief backward glance.

During the prolonged period of rapid world economic growth that began with reconstruction following World War II in Europe and Japan and continued until the 1973 oil crisis, Western Europe emerged as a modern industrial force and a major contributor to the driving force of world economic growth—international trade. The price of this rapid growth in prosperity—indeed to a considerable extent the major force permitting the growth to take place—was an increasing reliance on external sources of energy, in the form of oil. Europe's economies, principally coal-based since the time of the industrial revolution, found in oil a relatively cheap, plentiful, and highly desirable fuel—so much so that it was only through massive government subvention that the European coal industry managed to survive at all.

Figure 3-1 illustrates the fundamental nature of the change that the European energy economy underwent in the period 1950 to 1973. Oil consumption grew over the period at an annual rate of 12.1 percent, to occupy a 60-percent share of total primary-energy consumption in 1973, compared with a mere 12 percent in 1950. The lion's share of oil consumption was supplied by imports and OECD-Europe accounted for 57 percent of total OECD (Organization for Economic Cooperation and Development) oil imports in 1973. Europe consumed more oil in 1973 than it did of all primary energy forms in 1960.

Nor were the benefits of the postwar discoveries of vast new oil reserves, particularly in the Middle East, restricted to European countries. Japan built itself into the OECD's second largest economy, relying on imported oil for all but 5 percent of its energy in 1973. The United States exploited its own extensive resources of petroleum but still emerged as a large oil importer toward the end of this period. The industrialized world's rapid growth in oil consumption continued until the end of 1973 and created the market conditions under which the oil-producing countries within OPEC, aided by a political crisis, were able to quadruple the international price of crude oil at the end of that year.

This abrupt and massive increase in the price paid for a vital part of their energy supplies, coupled with the effects of the accompanying oil embargo, brought home to European countries the extent of their economic dependence on external supplies and the need to make efforts to reduce that dependence.

Reactions to the 1973-1974 Oil Crisis: The Creation of IEA

Another dimension of the crisis was that it became evident that the problem could not be solved by any one country, or even regional grouping, acting alone. Actions taken by any one country could have repercussions extending far beyond its frontiers and only by a higher degree of international cooperation could the most appropriate set of national responses to the crisis be worked out.

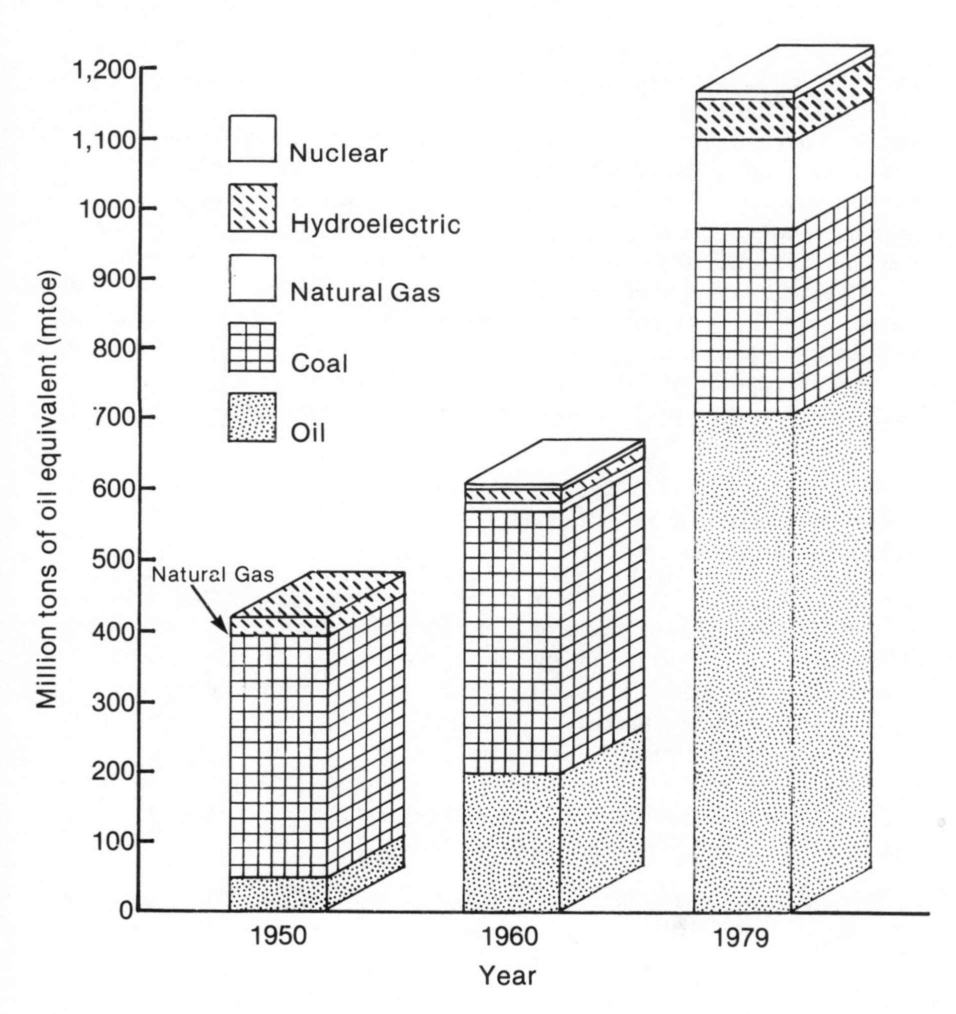

Figure 3-1. Historic Consumption of Different Forms of Energy in OECD-Europe

In fact, governments in industrialized countries first reacted in an uncoordinated way to the oil-supply interruption and selective embargo of 1973-1974, and it was the international oil companies that shouldered the responsibility of sharing available oil as equitably as possible. This task they carried out as well as they could under the circumstances, and it is significant that criticism of their role during the 1973-1974 crisis has been relatively muted—indeed, consuming-country governments generally recognized that the companies had played a valuable role in equalizing the misery and in avoiding some political difficulties that might have had serious international implications.

But—and this is perhaps the most significant consequence of that crisis—oil had become politicized. Governments became aware of their vulnerability to sudden interruption of oil supplies and their general responsibility for sharing of oil in times of supply interruption, for whatever reason. They also became aware of their longer-term economic vulnerability, stemming from overdependence on imported oil—their vulnerability to price increases, to political events, and eventually to physical limits of global oil supply. Finally, events of late 1973 brought home to governments the important role of energy in their increasing economic interdependence.

Thus, after a period of negotiation on a detailed program of international cooperation, the Agreement on an International Energy Program was signed and the International Energy Agency (IEA) created in November 1974, as an autonomous agency within the OECD. The IEA now has twenty-one member countries, Portugal being the most recent signatory to the Agreement in April 1980.

Objectives of the IEA and Progress during Its First Five Years

The general purpose of the IEA is to help coordinate government response of industrialized countries to global energy problems. This has both short- and longer-term aspects. In the short term the IEA exists to reduce vulnerability to sudden shocks by means of a workable emergency oil-sharing system, characterized by the automaticity of its activation and by the full cooperation of the oil industry.

The longer-term response involves a restructuring of the energy economies of the industrialized countries toward less heavy dependence on oil. There are three main dimensions of this effort:

1. The reduction of waste and the increase of efficiency of energy utilization.
2. An accelerated development of alternative-energy resources such as coal, nuclear power, natural gas, and renewable sources of energy.
3. The encouragement of energy research and development.

Basic to the entire effort is a more comprehensive understanding of the energy field and better information on energy. The IEA has devoted considerable attention to the improvement of data and the creation of new data systems, particularly in regard to oil production, consumption, imports, and prices. It has also been active in the other important aspect of information, namely an endeavor to provide the public with a better understanding of the energy problems and prospects it will face in the future.

There has been considerable progress during the first six years of the IEA's existence. The emergency oil-sharing system is now set up and can be made operational if necessary. Commitments on long-term cooperation have been made by the member countries, covering the three main dimensions of the long-term response already mentioned.

Energy efficiency has made some progress, as measured by the energy input per unit of Gross Domestic Product (GDP) in IEA countries. Before 1973, energy consumption rose at a slightly faster rate than did economic activity in general. Although statistics since 1973 cover a period of deep recession followed by a slow and hesitant recovery, the indications are that a significant uncoupling of energy and economic growth has occurred, and we expect energy consumption to grow over the coming decade at three-fifths or less of the general economic growth rate.

This has happened as a result of a higher energy-process stimulating investment in general energy saving (as well as measures taken by households and industries to reduce waste). It has been supported and stimulated by government-policy efforts in the field of energy conservation—financial or fiscal incentives, regulatory and informational campaigns foremost among them. October 1979 was designated International Energy Conservation Month by the IEA, and there were exhibitions, publicity campaigns, lectures, and conferences throughout the IEA area on the subject, bringing a greater public awareness of the need to save energy and ways to accomplish this.

On the energy-supply side, new-oil production in the IEA has commenced principally in the North Sea and Alaska, and exploration is increasing in most areas where good prospects of finding hydrocarbons exist.

Finally, energy R&D has been intensified. There are now fifty-two cooperative R&D projects, each involving several IEA countries, covering virtually the entire fields of energy use, energy conversion, and possible new sources of energy.

But on the other side of the balance, it is fair to say that world coal and natural gas remain underused, nuclear supply projections for the 1980s have been continuously reduced since 1973, and the global importance of oil to the industrialized economies has remained excessively high.

The main cause of this continued dependence on oil and the main obstacle to a transition away from oil use is that an economic structure and lifestyle based primarily on oil, introduced progressively over several decades, has a certain resistance to change; it is difficult to modify abruptly without creating severe social and economic problems. Efforts to begin the transition were not made any easier by the apparent surplus of oil during the years 1975-1978, which deemphasized the energy question and prevented democratically elected governments from taking strong action in this field.

Oil Supply and Demand: 1973-1980

The single most significant factor in Europe's oil situation in the latter half of the 1970s was the new supply from the North Sea. Oil production in Europe, almost negligible in the past, increased rapidly as the North Sea discoveries came onstream. It passed through the 2-million-barrels-per-day (mbd) mark during 1979. The peak of North Sea production and the shape of the subsequent decline will depend on the magnitude of future discoveries and on the depletion policies followed by the governments concerned, principally the United Kingdom and Norway. Production is predicted to reach between 3 and 3.5 mbd by 1983, and subsequently flatten out.

Largely because of this new indigenous petroleum supply, Europe's oil imports were much the same in 1978 as they had been in 1973, notwithstanding some growth in consumption. Imports had been falling as the North Sea production built up, even before the 1979 oil crisis. Oil-consumption growth in Europe had already slowed significantly in the post-1973 period, and although it picked up with the 1976-1979 economic recovery, the latest price developments are expected to have a significant impact on its growth rate in the future.

Forecasting oil consumption and supply in present circumstances probably involves more uncertainty than ever before. Today's environment is totally different from anything we have experienced in the past. This can perhaps be best illustrated by reference to classical techniques of energy forecasting.

The traditional method of forecasting oil demand involves starting from a projected economic growth rate; assuming a certain energy/GDP elasticity to project energy consumption, estimating the supply of energy from non-oil sources, and arriving finally at an oil-demand figure. The progress was then taken a step further by deducting from this the estimated supply from non-OPEC sources to arrive at a figure for assumed OPEC production.

The soundness of such a technique is now, to say the least, questionable. Future oil consumption (assuming the avoidance of a deep global economic depression) may well be supply-constrained throughout the 1980s. These constraints may well pass through other energy-source limitations to economic growth itself, but it is still too early to make quantitative assertions about this.

The post-1973 development of European oil production and consumption, and overall energy consumption, is shown in figure 3-2. Part of the slowing of oil growth has been an increasing contribution from other fuels. Oil's share in total primary energy has been decreasing from 60 percent in 1973 to 56 percent in 1978, and is likely to continue to fall over the coming decade.

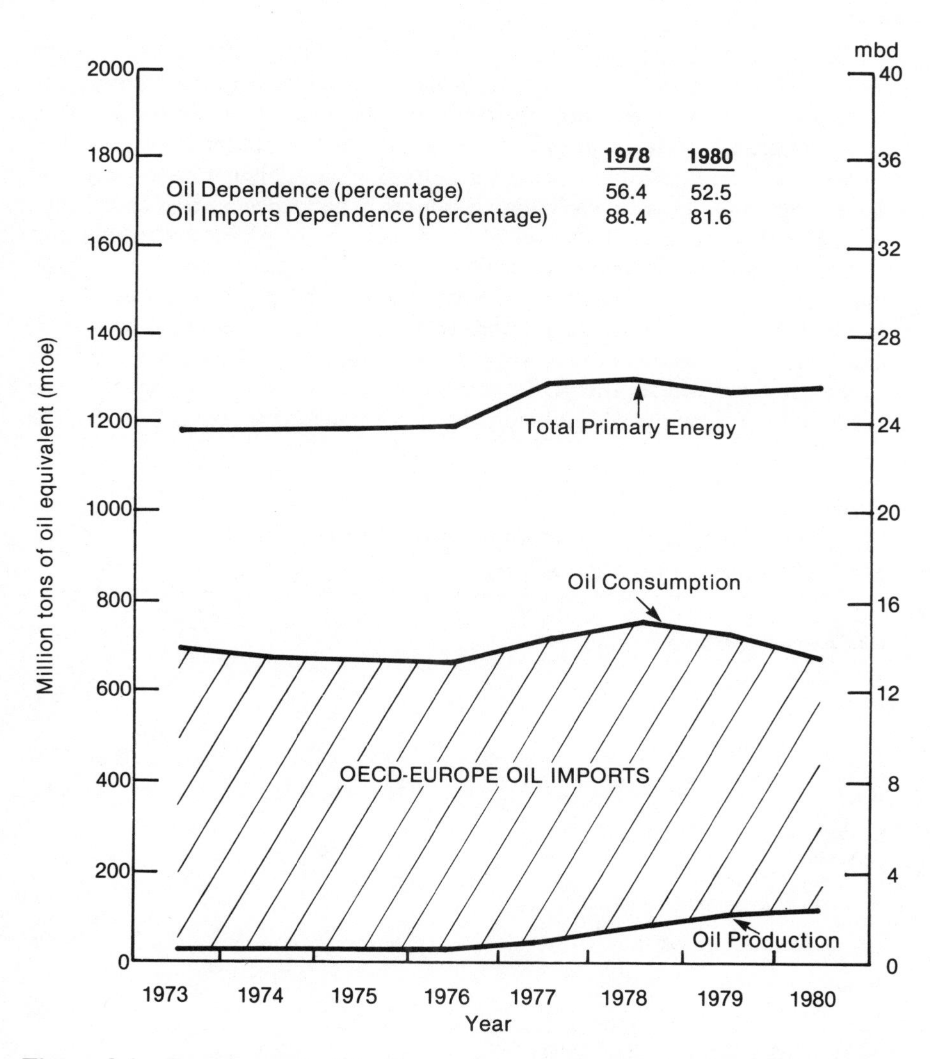

Figure 3-2. Energy and Oil Demand and Supply in OECD-Europe, 1973-1980

It is perhaps misleading to regard Europe as a unit, considering the divergent trends in energy dependence that the arrival onstream of North Sea oil has caused. From being 97 percent dependent on imports for its oil supply before the arrival of North Sea oil, OECD-Europe's average external dependence has fallen to a little over 80 percent in 1980. But most European countries will still be importing more than 95 percent of their oil needs throughout the 1980s. They will benefit from Europe's indigenous oil supply only through its impact on the world market.

The events of early 1979 have demonstrated clearly that at a time when oil-supply prospects have swung rapidly from surplus to stringency, the need for a cooperative approach to the energy problem is reinforced if all countries are not to be made worse off. If oil supply is limited it becomes imperative to avoid an international scramble for oil. This is particularly true for Europe with its wide variation of import-dependence among countries.

Table 3-1 places OECD-Europe in the context of global oil supply and demand in millions of barrels per day.[1] OECD oil consumption was much the same in 1979 as it was in 1978, and it declined by 2.5 million barrels per day in 1980 as a result of high oil prices, efforts by governments to curb oil demand, and slower economic growth. Building a 1985 scenario around the Tokyo Summit and IEA import commitments would show oil consumption in the industrialized world growing very slowly, giving an overall 1978-1985

Table 3-1
World Oil Consumption and Supply, 1979-1980[a]
millions of barrels per day (mbd)

	Year	
Oil Consumption/Supply	*1979*	*1980*
Oil consumption		
IEA	38.7	36.4
OECD	41.5	39.0
(of which Europe)	15.6	15.1
Developing countries	10.2	10.3
World	51.7	49.3
Oil Supply		
IEA/OECD	14.1	15.2
(of which Europe)	2.4	2.9
Non-OPEC developing countries	5.2	5.8
OPEC	31.6	27.6
Net imports from CPEs[b]	1.1	1.0
Total	52.8	49.6
Balance	0.2	0.8

Source: International Energy Agency, Paris.

[a]Including natural gas liquids.

[b]Centrally Planned Economy Countries (USSR, China, Eastern Europe).

oil growth rate of about 1 percent per annum. More recent developments in OPEC oil-supply potential, in world economic-growth prospects, and in the demand response to oil-price increases and policy measures to reduce oil consumption, would indicate that global oil supply and demand levels in 1985 may show little change compared with those in 1980. There may be some decrease in the developed world's oil consumption because of increased efficiency of use and fuel switching, balanced by an increase in consumption in the developing world, particularly in OPEC countries.

The Role of the Oil Industry

The oil industry, from its birth in the last century, has been characterized by structural evolution and adaptation to changing circumstances. This is of course true of any successful industry. There were some very significant changes in the period between 1970 and 1978: the ownership of oil (through concessions in geologically prospective areas) passed increasingly from oil companies into the hands of the producing countries themselves. These countries in some cases nationalized concessions outright, and in others increased their participation gradually until they exercised majority or complete control over their oil production. The system of concessions has still not completely disappeared, but by far the bulk of internationally traded oil is now purchased from producing-country governments or state oil companies.

This gradual change of ownership did not significantly affect the channels through which oil flowed from the well to the consumer until 1979. Although the international oil companies no longer *owned* the concessions, they retained their rights as primary off-takers, buying the crude oil from the producing country and disposing of it as before, either in their own affiliate system or through sales to third parties. The major companies suffered some gradual reduction in the overall amounts of crude oil they handled as primary suppliers, as smaller independent oil companies increased their international activities, as national oil companies of producing countries became more active, and as governments of some importing nations began to purchase crude oil directly in state-to-state deals involving money, goods, or technology in exchange.

But the net result was a quite minor perturbation on the overall market situation, until the events in Iran early in 1979. The crisis of 1979 began a new phase of restructuring of the oil-supply system that may have far-reaching consequences for the future.

All major oil companies obtained supplies from Iran through the Consortium (formed in 1954 following nationalization; included British Petroleum, Royal Dutch/Shell, and a number of other companies), so all were affected by the cut-off (to extents varying from 5 percent to 45 percent of their overall crude supplies). Moreover, because of interlocking supply and sales arrangements, this disturbance in one corner of the worldwide crude-oil-supply system spread rapidly through the entire system like a ripple in a pond. Just to take one example, British Petroleum (BP) used to sell 360,000 barrels/day to Exxon and 225,000 barrels/day to Petrofina. BP lost more than 1 mbd supply from Iran, so was obliged to declare *force majeure* on these sales, which in turn led Exxon and Petrofina to declare *force majeure* on their customers; Petrofina cut 35 percent from its sales of North Sea crude to third parties.

Although Iran began exporting oil again in the spring of 1979, albeit for smaller quantities than before, the patern of purchasing of Iranian oil was entirely changed. The Consortium was to all intents and purposes defunct, and instead of ten Consortium members with stable shares, there were almost forty companies—majors, independents, refining companies, and state oil companies—sharing nine-month contracts for less than half the previous export volume purchased directly from the National Iranian Oil Company. These companies had no assurance that their contracts would be renewed in 1980, and very little influence on the conditions of renewal. Indeed, the cessation of exports occasioned by the outbreak of war between Iran and Iraq demonstrated clearly the fragility of all supply contracts.

The tight oil market that persisted throughout 1979 resulted in the maintenance of high spot-market prices for crude oil and oil products. This was partly because the margin, or "buffer" of spare production capacity in OPEC countries, thought to amount to several million barrels per day as recently as 1978, no longer existed. But it also owed its origins to the industry's experience in Iran, an experience that sent the major companies and those who traditionally bought oil from them searching, sometimes in competition with each other, for new supplies elsewhere.

In turn, the persistence of this multitier system of oil prices in the market, under which prices at one time in mid-1979 ranged from under $18 per barrel for Saudi Light marker crude to higher than $40 per barrel, encouraged other producing countries than Iran, notably Kuwait, Nigeria, and Libya, to withhold some of their oil previously contracted at the official selling price, for sale to new clients at spot-related prices. This shift brought with it a trend toward shorter and less rigid contracts in many producing countries. Although information on exactly how much crude oil was being sold in mid-1979 at prices higher than OPEC official selling prices is hard to obtain, a rough estimate might be 10 to 15 percent of OPEC production.

Thus, a reshuffle of the international crude-oil distribution system took place during 1979, with the major companies losing access to about 1 mbd compared with 1978, even though overall OPEC production increased about 1 mbd. A large number of purchasing entities moved to the upstream end of the market to buy crude oil. Faced with a fall in the net availability of crude oil, the international majors reduced substantially their traditional role as net sellers of crude oil and withdrew largely from what is termed the third-party market in order to supply fully their own affiliate systems.

Although the overall balance of supply and demand improved during the second half of 1979, this contributed less to calming the market than might be expected. The majors continued to lose direct access to crude oil at the wellhead, to the extent of a further 1 mbd in the first quarter of 1980. Subjected to continuing change at the upstream end of the system, uncertainty motivated individual buying behavior more than the global balance between supply and demand.

Intimately related to uncertainty is loss of flexibility. No single oil company can balance within its own system its crude oil availability and its refining and marketing needs, and will always need to trade some crude oil and/or products. But it is easier to find this balance within a larger unit, and the majors have traditionally had a strong balancing and stabilizing function in the market.

The segmentation of buyers and decreased flexibility also implied the need for greater stocks in aggregate, as all participants in the market sought to achieve what they perceived to be an adequate level of security, this level itself increased by the high degree of uncertainty as to future supplies.

It is through a combination of these influences that a supply/demand picture that looks globally satisfactory, when followed through to oil-industry participants in the market, can still leave a high degree of insecurity, with all of the implications this can have for oil prices.

This is something of a vicious circle. It can be broken only when a margin of supply capacity builds up through reduced oil consumption, and the distribution system settles into a more stable configuration, thereby lowering the level of uncertainty among market participants. This safety margin seemed a likely prospect earlier in 1980, but the outbreak of hostilities between Iran and Iraq again raised the level of uncertainty concerning supplies.

These rather substantial changes in the industry's distribution system were without doubt the most significant influences on the worldwide oil market in 1979 and early 1980. They have been explained in greater detail and set in their historical context elsewhere in an analytical paper by the author of this chapter.[2]

Europe has certainly not escaped the disturbances of 1979 in the oil market. A number of European countries had traditionally relied on the spot market, centered in Rotterdam, for a good part of their supplies of oil products.[3] Over the last few years of oil surplus these were low-cost supplies, but in 1979 these countries saw the cost of this element of their product supplies more than double in the space of a few months, as surplus turned to shortage. Countries that relied primarily on long-term contracts from major oil companies or from their own state companies were less hard-hit, but their supplies were still to some extent cut by *force majeure* clauses in their supply contracts, and they also had to seek some new, generally higher-priced supplies.

Some European countries, notably France and Italy, intensified their efforts to obtain a larger proportion of their oil supplies directly from producing countries through government intervention. In fact, all countries' governments became more involved in the security of supply aspect. The growth of state-to-state deals testifies to this, especially in the period after the 1979 crisis. The amount of crude oil involved in the various types of government-government deals increased by some 50 percent in 1979 compared with 1978, from under 4 mbd to about 6 mbd. European countries accounted for about half of this increase.

That Europe should show such a trend is no surprise. Among the industrialized-country groupings, Europe is perhaps the most vulnerable to energy-supply problems. Although Japan's relative dependence on oil is greater than Europe's, Japan is pursuing a homogeneous and in-the-main successful policy of diversifying external supplies, especially since the events of 1979 demonstrated the vulnerability associated with relying on a few large international company sellers.

The disparities among European nations in energy and oil self-sufficiency, added to the wide variation of economic strength, are divisive influences, especially in a world of competition for scarce supplies. Happily, the lack of cohesion that had been a feature of European energy policies in recent years appears to be giving way to a new, more cooperative approach in the context of the European Communities (EC).

In summary, oil-industry structure in Europe over the coming decade will be shaped by the general political desire for security of supply. This probably means more government participation, either directly through state oil companies or indirectly through close contact and information flow between governments and oil companies. Within Europe, the Communities are the focal point for coordination of these policies, and in looking outwards, the IEA provides the forum for a wider degree of cooperation among the industrialized countries.

Role of the IEA in the 1980s

The objectives and activities of the IEA have already been described. Faced with a period of supply stringency that may well last, with minor variations, throughout the next decade, how can the IEA play a constructive and valuable role?

Although the transition from primary reliance on oil has already begun, this fuel will in all probability remain at the center of the world's energy stage at least through the 1980s.

In the oil sector, the IEA has set import targets for 1985 and monitors progress continuously so that these targets will not be exceeded. A part of this work, which underlies but extends beyond the targetry, is the need to develop and maintain improved information systems. The IEA has now developed a comprehensive registration procedure covering all oil imports into member countries: price, volume, and other relevant details that expand its data system very significantly. With these in place, the IEA countries and the IEA headquarters in Paris have a much better coverage of the range of prices at which crude oil and its products are sold to member countries.

An important aspect of the IEA's work is to improve coordination among member countries in times of supply stringency of a nonemergency type. Basically this means trying to avoid a worldwide scramble for oil to the detriment of all. The events of 1979 have amply demonstrated the magnitude of problems that can be caused in pricing by what is really an oil shortage of only a small percentage.

Part of this task is to keep a close watch on possible sources of friction in the oil market. Industry structural changes and spot-market behavior have already been mentioned. Oil-stock levels in member countries need to be closely monitored, as stock changes can add to instability in a tight market situation. IEA ministers have agreed to coordinate stock policies in an effort to minimize this possible source of market pressure. Also of importance is the flexibility of the worldwide refining system to meet changing demand for products. And looking toward the longer term, the amount of exploration for hydrocarbons and enhanced recovery—problems associated with these and the possible ways governments might be involved in encouraging them—are topics the IEA will have to keep under close review. The oil that will be consumed in the last years of this century will have to be discovered in the near future.

In all this work, the tradition of close contact with the oil industry, built up since the IEA's creation, will need to be maintained, since it is the industry that will continue to find the oil and physically move it to the consumer. Governments may make the decisions regarding the conditions under which the oil is moved within their jurisdictions, but they possess neither the technology nor the logistical expertise that the industry has built up over many decades of experience.

Because the IEA is not solely concerned with oil, and because success in its other fields of activity will influence the oil sector in the long run, it is appropriate to describe briefly the activities in other sectors that will concern the agency in the decade to come.

Energy conservation is a very important field. As noted before, progress is being made. It needs to be encouraged, first by ensuring that energy prices reflect real world-market levels in all countries, then by adding government policy incentives, because wasting energy has social and international strategic costs that are not directly reflected by the market. The IEA is engaged in an in-depth analysis of energy consumption patterns and their relationship to economic activity, in an attempt to pinpoint areas where maximum progress in increasing energy efficiency may be realized.

The IEA is also active in encouraging, through the polices of its member countries, the development of alternative energy resources. It is coordinating policies on expanding international coal production, use, and trade between countries. It is devoting considerable attention to the problems and prospects of nuclear power after the Three-Mile island accident: the need to formulate and adopt appropriate safety standards to avoid such accidents in the future; to resolve long-term waste-disposal questions; and to ensure an adequate international supply of nuclear fuel and equipment. It is also studying the prospects for development of natural-gas supply and international trade.

Energy R&D is being continuously expanded through cooperative projects involving IEA countries. The R&D program has now been opened to participation by nonmember countries, including developing countries. An international energy-systems analysis project, looking up to fifty years ahead, has now been completed and provides an analytical tool for the development of energy strategies by the IEA and its member countries.

Lastly, since democratic governments need the support of the voters for the strong action that will be needed in the energy field, the IEA is devoting an increasing amount of attention to public information and persuasion, in parallel with its member governments. If the man in the street continues to believe that there is no energy problem, or that it is all a conspiracy by the oil companies to increase their profits, or that oil tankers are lined up out at sea waiting for the price to go up, or that burnoosed Arab sheikhs are engaged in a coordinated international rip-off to take away from us our hard-earned prosperity, then the cure for our addiction to oil will be delayed until a really serious crisis causes more permanent damage to our economies, to our way of life, and perhaps to our institutions of government.

Notes

1. These energy balances exclude the centrally-planned-economy countries (CPEs), specifically the USSR, China, and Eastern European countries.

2. Ian M. Torrens, "Changing Structures in the World Oil Market," *The Atlantic Papers* 41 (Paris: The Atlantic Institute for International Affairs, 1980).

3. This consists of single cargo or barge (even tank-truck) sales, and has traditionally been part of the balancing mechanism for marginal quantities of oil, both crude and oil products. Though not located physically at Rotterdam (it consists mainly of telephone or telex links between buyers and sellers scattered throughout the world) the presence of large refining and storage facilities in the Rotterdam area makes it an important center for international cargo trading.

4 Nuclear Power, Trade, and Nuclear Nonproliferation

Georges Delcoigne

World energy needs are growing rapidly and it is expected that the present world consumption of about 6 billion tons of oil equivalent will increase to more than twice this level by the year 2000, to 12 to 14 billion tons of oil equivalent.

By that time, the population of the world, which is of the order of 4.5 billion people today, will increase to over 6 billion. Most of this growth will take place in developing countries which, by then, will account for more than two-thirds of the total. There is a striking disparity in the present distribution of energy consumption: industrial countries' energy consumption is more than ten times that of developing nations. The initial stages of development are particularly energy-intensive, so as the developing countries progress there will necessarily be a rapid growth in energy demand. Thus, even if by a miracle of active conservation and restrictive policies the industrialized nations were to maintain zero energy growth, the pressure of demand from developing countries would bring about a substantial increase in world energy needs.

Energy Supplies

What are the potential sources of supply on which mankind can draw to meet this doubling of energy demands? According to conservative estimates by the World Energy Conference, the cumulative energy consumption of the world over the next twenty years would be of the order of 250 billion tons of oil equivalent.

> For *oil* to maintain its share of the market would require a cumulative production of more than 110 billion tons: present proven reserves are 90 billion tons.
>
> Resources of *coal* probably exceed those of oil and gas by an order of magnitude, yet they share with hydrocarbons the serious shortcoming of highly uneven distribution among the various countries of the world.

The opinions expressed in this chapter are those of the author and do not necessarily reflect those of the organization for which he works.

Expansion of coal production also gives rise to serious environmental and social problems.

Renewable sources of energy must be developed at maximum speed, but they are available in diffuse and irregular forms (solar and wind power) and offer only partial solutions. Their combined share of world supply is unlikely to exceed 10 percent (at the most) of the total by the year 2000. The present figure amounts to about 1 percent of the total.

Nuclear power. There were at the end of 1980, 271 nuclear power reactors in operation in the world with a generating capacity of over 149,000 megawatts of electricity (MWe) or about 10 percent of the world's total electrical generating capacity. There are 230 more reactors under construction or on order, representing some 211,000 MWe. Present plans call for 454 nuclear-power reactors to be in operation by 1985. Some countries already derive a major share of their electricity supply from nuclear power. In Western Europe, for instance, Belgium, Sweden, and Switzerland obtain over 25 percent of their electricity from nuclear power, and by 1985 this percentage will be of the order of 40 to 50 percent.

Table 4-1 shows total electricity generation in megawatts of electricity (MWe) and the percentage of nuclear-electricity consumption in the United States, Canada, Western European countries, Japan, and some developing nations. Figures are for June 1980.

Nuclear power is technologically and commercially ripe for expansion. Present known uranium resources, up to a cost of $130 per kilogram (kg) of uranium, are 2.6 million tons. Adding another 2.5 million tons of estimated additional reserves, these resources would cover the needs of the maximum nuclear power programs up to the year 2010. (The search for uranium has so far been concentrated on low-cost deposits in selected countries, and has not yet touched upon wide areas of the world such as Latin America and Southeast Asia.) In the longer term, nuclear power based on breeder reactors is the only energy form depending more on human skills than on natural resources.

The United States

The United States is the world's largest single consumer (25 percent of world total), importer, and producer of energy. It also has a wider range of energy options than most other industrial countries. In addition, U.S. energy policies have profound repercussions in the rest of the world and particularly in Western Europe.

Table 4-1
Electricity Generation and Nuclear-Electricity Consumption in the World, 1981
Megawatts of electricity (MWe)

Country	Operating		Under Construction		On Order/ Planned		Percentage of Total Electricity Supplied by Nuclear Plants during 1979
	Number of Units	Total MWe	Number of Units	Total MWe	Number of Units	Total MWe	
CECD Europe							
Belgium	3	1,664	4	3,807	—	—	22.0
Finland	4	2,160	—	—	1	1,000	19.0
France	28	19,885	24	25,710	22	27,600	18.0
Federal Republic of Germany	14	8,606	10	10,636	13	16,127	11.0
Italy	4	1,417	3	1,999	2	1,930	1.5
Netherlands	2	501	—	—	—	—	6.0
Spain	4	1,973	10	9,152	5	5,015	6.0
Sweden	9	6,415	3	3,025	—	—	23.0
Switzerland	4	1,940	1	942	—	—	26.0
United Kingdom	31	7,002	10	6,158	1	1,100	13.0
OECD North America							
Canada	11	5,494	14	9,751	—	—	11.0
United States	73	54,680	83	91,340	21	23,663	12.0
OECD Pacific							
Japan	24	14,994	11	9,127	3	3,017	10.0
CMEA							
Bulgaria	3	1,224	2	1,408	3	3,000	19.0
Cuba	2	800	1	408	1	408	—
Czechoslovakia	2	801	6	2,520	13	9,316	2.5
German Democratic Republic	5	1,694	4	1,644	6	2,448	9.0[a]
Hungary	—	—	2	816	2	816	—
Soviet Union (USSR)	35	14,036	17	16,260	26	26,600	4.0[a]
Poland	—	—	—	—	4	1,632	—
Romania	—	—	1	660	2	1,068	—

Table 4-1 *(continued)*

Country	*Operating*		*Under Construction*		*On Order/ Planned*		*Percentage of Total Electricity Supplied by Nuclear Plants during 1979*
	Number of Units	*Total MWe*	*Number of Units*	*Total MWe*	*Number of Units*	*Total MWe*	
Developing Countries							
Argentina	1	335	2	1,292	3	1,800	7.5
Brazil	—	—	3	3,116	—	—	—
Egypt	—	—	—	—	1	622	—
India	4	809	4	880	—	—	2.5
Republic of Korea	1	564	6	4,967	2	1,900	8.0[a]
Mexico	—	—	2	1,308	—	—	—
Pakistan	1	125	—	—	1	600	0.3
Philippines	—	—	1	620	—	—	—
Thailand	—	—	—	—	1	600	—
Turkey	—	—	—	—	1	672	—
Yugoslavia	—	—	1	632	1	1,000	—
Others							
Israel	—	—	—	—	1	600	—
South Africa	—	—	2	1,843	—	—	—
Taiwan	3	2,159	3	2,765	—	—	17.0[a]
Total	268	135,278	230	212,788	136	132,534	10.0[b]

Source: *Nuclear Power Reactors in the World, Ref. Data Series No. 2, 1981 Edition* (Vienna: IAEA, September 1981); *Power Reactors in Member States, 1981 Edition* (Vienna: IAEA, June 1980); *Power Reactors in Member States, 1980 Edition* (Vienna: IAEA, June 1980).

Note: Nuclear programmes in Austria and in Iran have been interrupted and the plants are not included.

[a]Estimated.

[b]Average for countries with nuclear plants operating in 1979.

Total U.S. energy consumption today is the equivalent of about 37 million barrels per day of oil. Of that, 27 million are derived from domestic production. Over a period of ten years or so, that 27 million barrels could perhaps be stretched to 32 million. But during that period, overall consumption would almost surely rise, so even the increase of 5 million barrels daily would not ease the problem of imports.[1]

The greatest single matter of concern posed by increasing U.S. oil imports is its effect on world oil prices: the higher the price of U.S. oil imports, the higher will be world oil prices, and this dependence is destabilizing. The role of U.S. imports is crucial. If the United States does not act to dampen imports, Europe and Japan will find it harder to restrain their own demand.

There are four conventional sources of domestic energy: oil, natural gas, coal, and nuclear power. However, the outlook is bleak. All four are likely to deliver less energy than projections by advocates would lead one to believe.[3]

For oil, no matter what happens to domestic prices or to the structure of the U.S. oil industry, the fact remains that the physical production of oil from conventional sources will continue to decline. For natural gas the best that can be expected is that a deregulated price will enable production to remain at current levels.

Domestically produced coal, according to President Carter's National Energy Plan, should partly replace imported oil. To obtain a large increase in output, a traditionally backward industry must quickly be transformed into a modern, technologically advanced one. In addition, potential users are reluctant to commit themselves to coal, because of the uncertainty of meeting future, and as yet unspecified, environmental requirements. It has been calculated, for instance, that if a tripling of U.S. production from now until the year 2000 was considered necessary to meet domestic and export demands, this would require 700 new mines; 340,000 additional mining labor force, and $116 billion in investment. This does not take into account coal-related transport facilities such as trains, barges, trucks, and slurry pipelines.

The U.S. production of electricity of nuclear origin in 1978 reached the level of 260 Terra Watt hours (TWh), 11 percent of total electricity consumption. However, further development of nuclear power is hampered by controversy. It has become a major political issue, and there have been long delays in the courts. The existing regulatory requirements are a further complication. The number of quality-assurance and quality-control standards applicable in the United States grew from about 100 in 1970 to about 1,600 in 1976. These regulatory requirements have doubled the amounts of many important commodities required (concrete, steel, pipes, cables, and so on), and about twice as many engineers are required because of the longer times per project. The confusion from the growth of regulatory requirements and procedures has complicated and caused delays in the procurement, engineering, and construction processes. Much of the increased cost has resulted from the procedures necessary to record and demonstrate compliance. As a result, to build a nuclear power plant in the United States now takes twelve to fourteen years. In France it takes approximately six years.

The Three-Mile Island accident of 1979 did nothing to simplify the situation. Following a comprehensive study and investigation of the accident, the report issued by the U.S. presidential commission chaired by John G. Kemeny is remarkable for its clear presentation of the events and its objectivity. It recommends steps to be taken in order to prevent a repetition of the accident. The Kemeny Commission's most important conclusions are that the equipment worked better than expected, but the operators did not. Or, in the commission's words: "It became clear that the fundamental problems are people-related problems."[3] The main factor that turned minor events into a potentially major catastrophe was human error, not mechanical failure. This was confirmed in 1980 by all forty-four nations with nuclear power programs that reviewed nuclear safety issues in Stockholm and concluded that there were no factors related to safety that limit the use and development of nuclear power.

The United States still expects a growing contribution from nuclear power for the late 1980s from the plants that are now under construction. In 1981 it had seventy-three nuclear power reactors in operation.

Western Europe

Western European oil imports amount to well over 50 percent of European energy needs. The impact of the oil crisis of 1973 on both the supply of energy and the general economic situation still makes it difficult to predict with any degree of confidence the growth of nuclear energy production.

But in Europe since 1978 the message from European politicians and institutions has been, bluntly, to develop coal and nuclear power before oil supplies begin to run out. A series of studies and pronouncements by the European Communities have stressed the need for a significant growth in nuclear capacity to ensure economic growth in the coming decades. Adopted in 1980, EC energy objectives for 1990 include an expansion of the role of nuclear power, along with coal, in the production of electricity. The Commission has noted that "Nuclear Power offers a production cost advantage over coal and oil for base load electricity generation."[4]

This view is also held by organizations with a wider membership, such as the International Energy Agency (IEA), in Paris. Ministers of its member nations agreed in June 1979 on the need for projected additions to nuclear power supply, to be exceeded wherever possible, and the final communiqué of the Tokyo Summit (end of June 1979) declared that: "Without the expansion of nuclear power generating capacity in the coming decades, economic growth and higher employment will be hard to achieve." But some countries are awaiting the impact of energy conservation measures. In others, public concern over the risks of nuclear power has led to burgeoning licensing requirements and to de facto moratoria on nuclear power plant construction. Nevertheless, in June 1980 the ministers and heads of state agreed at the Venice Economic Summit to ambitious objectives for 1990. They agreed to take action to accelerate structural changes in their energy economies. Agreement was reached that non-oil fuels should contribute an additional 15 to 20 million barrels of oil equivalent (mboe) by 1990. This would require a doubling of coal power and a 2.5-fold increase of nuclear power.[5]

The Federal Republic of Germany has fourteen operating nuclear plants, and there are plans for eleven more to come into operation by about 1985. More stringent design and licensing requirements have increased capital investment costs by 15 to 20 percent per year.

Moreover, during the political campaigning that led to the 1980 elections, the appearance of a stridently antinuclear political party, the Greens (who ultimately received only 1.5 percent of the vote), threatened to upset the political balance in the Federal Republic and contributed to a political standstill in nuclear-power development.

In France, nuclear-powered electricity will provide about 50 percent of the total electricity production by the year 1985. This will be achieved by building some 35,000 MWe of PWR (Pressurized-Water Reactor) stations plus a 1,200 MWe fast-breeder-reactor station in the fifteen years from the start of the program in 1970. There is a clear economic and political strategy to secure energy independence. Cost increases have been held down and the program is only slightly behind schedule. The French regulatory and licensing procedure is no less vigorous than in other countries but there are clear benefits from the high level of standardization and from the fact that there are no intermediate licensing steps between the granting of the construction license and the operating license. Although existing institutions right up to Parliament have been involved in the licensing process, the type of procedural tactics employed in German and American courts to delay the development of nuclear energy are not tolerated by the due process of law in France.

In Sweden, with its desire to become independent from imported oil and its decision to halt further expansion of hydro power for environmental reasons, nuclear power was introduced at an early stage in order to use its large uranium deposits. The nuclear program of the early 1970s foresaw nuclear electricity meeting 40 percent of the total demand by 1985 from thirteen power units. Twelve of these have so far been built or are under construction; they produce some 25 percent of Sweden's electricity. Following a growing political controversy over the use of nuclear power, the Swedish government accepted in 1978 the recommendation of an Energy Commission for the completion and operation of all nuclear power plants under construction, but decided the program should be halted at that point with twelve units operating. The Three-Mile Island accident led the Socialist party to reverse its previous stand and support a national referendum to decide the future of Sweden's nuclear industry in March 1980. The decision of 58 percent of the Swedish voters was to complete their planned program and load the four completed plants that had lain idle, pending the outcome of the referendum.

In Switzerland, 75 percent of the energy supply has been met by imported oil, and one result of the oil crisis is a more realistic attitude toward nuclear power, even after the Three-Mile Island accident. Nuclear electricity accounted for over 30 percent of total electricity production in 1980. Swiss electricity should be 50 percent nuclear by 1981, when the power station at Leibstadt comes into operation.

In the United Kingdom, despite the rapid development of North Sea oil, a state of energy self-sufficiency will only exist for a few years during the second half of the 1980s; thereafter energy imports will be resorted to on an increasing scale. The British government is now proposing an expanded program by which 50 percent of electricity needs will be met by nuclear power in the year 2000. The United Kingdom also has a 250 MW fast breeder in operation, and plans to build a large commercial demonstration plant are to be considered.

Japan and Eastern Europe

In order to complete the picture, it should be noted that Japan, a country with little indigenous energy and few fuel resources (only about 10 percent of the total), has no choice but nuclear energy. It is now the country with the second largest nuclear capacity in the world, next to the United States. In addition to twenty-four operating reactors, another fourteen are under construction or under preparation for construction. Japan's total nuclear capacity in 1985 should be 28,000 MWe, supplied by thirty-five nuclear units. The capacity target for 1990 is 50,000 MWe, or 22 percent of its electricity.

The COMECON (Council for Mutual Economic Assistance) Soviet-bloc countries also face problems in their energy requirements. Because the Soviet Union (USSR) is the major supplier of oil to COMECON members and the growth of Soviet oil production is already slowing because of shortfalls, the USSR and the Eastern European countries reached the conclusion that commercial nuclear power plants were the only key to a "limitless source of energy." As a result, the Eastern European COMECON countries signed an agreement in Moscow (June 28, 1979) committing themselves to install 150,000 MWe of nuclear power by 1990, which means that nuclear energy would provide about 30 percent of their electricity by then.[6]

Nuclear Trade

Until 1972, the United States filled 85 percent of the world orders for nuclear power plants and 90 percent of the enrichment services for the free world; by 1980 the Soviet Union had taken about one-third of the market. As a result of the restrictive export policies initiated by the Carter Administration and the differing U.S.-European emphasis on nuclear proliferation and other energy and security matters, the United States is no longer considered a reliable supplier of nuclear material by some of its allies and other trading partners. The European nuclear industry (France, West Germany, Great Britain, and Sweden) is also looking for export markets that would enable its capacity to be used outside its national frontiers. As the European countries develop their programs, they try to achieve self-sufficiency both in enrichment and in reprocessing.

The supply of enriched uranium is necessary for fueling the commonly used LWR (Light-Water Reactor) types, but enrichment is a difficult technology and the process has been kept secret. Because of its extremely high cost only a few countries can afford it. Commercial enrichment services are offered by two mixed government-private transnational consortia: URENCO (United Kingdom, Netherlands, and the Federal Republic of Germany), and EURODIF (Belgium, France, Italy, and Spain). Japan is planning to build a centrifuge plant. These facilities will provide a substantial degree of sufficiency. The trend away from U.S. enrichment services seems to be continuing: eight U.S. enrichment-services contracts have been cancelled by foreign customers since June 1978, often in favor of more expensive—but more reliable—contracts from other sources.[7]

For reprocessing there are at present commercial-size plants only in Great Britain and France. France and the Federal Republic of Germany, primarily, began considering negotiating sales that include reprocessing and enrichment facilities, to Brazil, Iran, Pakistan, and South Korea. In view of the growing use of nuclear power and steadily rising prices of uranium, France, West Germany, and the United Kingdom also increased their research in breeder reactors and advanced converter reactors (which could only be developed when reprocessing facilities exist).

This trend toward nuclear self-sufficiency was reinforced by the growing interest of some developing countries in minimizing the impact of the oil crisis through the acquisition of nuclear technology and the achievement of self-sufficiency in the nuclear fuel cycle. Within this group one could include Argentina, Brazil, Korea, Pakistan, South Africa, and Taiwan. Some of the oil-producing countries have also expressed an interest in the acquisition of nuclear technology (Algeria, Egypt, Iran, Iraq, Kuwait, and Libya).

These novel commercial deals in nuclear trade, which deal with sensitive technologies such as the arrangements for the sale of reprocessing plants to Brazil, Pakistan, and South Korea, in the long run favor the increase of self-sufficient states with a nuclear-weapons capability and add to the ongoing concern about nuclear weapons proliferation.

Commercial Nuclear Power and Nuclear Proliferation

It is interesting to see how the world arrived at the present nuclear situation. Though the potential for abusing peaceful nuclear technology is undeniable, *there has been no case* where a country has used its commercial nuclear power plants for the production of nuclear explosives. To emphasize this point: during the decade 1945 to 1954, three countries (the United States, Soviet Union, and Great Britain) detonated nuclear explosives. In the next decade, 1955 to 1964, another two countries (France and the People's Republic of China) and in the third decade, 1965 to 1974, one single country (India) did so. The rate of nuclear proliferation thus slowed down at the same time that nuclear technology and nuclear power were rapidly spreading. There were 5 MWe of nuclear power in operation in 1954, 9,000 MWe in 1968, and over 110,000 MWe in 1978. One can say that in 1945 only two or three nations had the technological know-how to make atomic bombs. Today this number is much larger—yet only six have exploited this capability.

Historically, therefore, there is no relationship between the expansion of nuclear power and the development of nuclear explosives. In fact, the two technologies have increasingly diverged, as there are cheaper and more expeditious ways of producing nuclear explosives than nuclear power plants.

In order to make weapons-quality material, all that is needed is natural uranium fuel, a research reactor, a moderator (which might be industrial-grade graphite or heavy water), and a simple reprocessing plant. American studies have estimated the total cost of the plant required at about $70 million. The infrastructure needed is that of one large nuclear research project. The alternative route is through a small high-separation enrichment plant.

In contrast, a nuclear power plant of today's standard size involves a capital investment of the order of $1,000 million—twenty times as much as the minimum weapons-material complex. It also requires the creation of a complicated administrative and regulatory infrastructure, the training of professional and specialist cadres, and for most countries except the major industrial powers, the plant and its fuel would have to be imported and in addition placed under international safeguards. No country wishing to establish quickly a nuclear-weapons capability would logically take this route. Most European countries, therefore, consider horizontal proliferation a problem of security and politics that is only exacerbated by export restrictions on peaceful nuclear technology.

Evolution of Safeguards and the Non-Proliferation Treaty (NPT)

After the second NPT (Non-Proliferation Treaty) Review Conference in 1980 and the end of the International Fuel Cycle Evaluation (INFCE), one should review the evolution of political applications of international safeguards attached to the development of the nuclear industry.

For as long as there has been international nuclear trade, supplier nations have demanded some kind of assurance that their supplies would not be used for military purposes, and particularly for the manufacture of nuclear weapons. To verify this they have demanded the application of safeguards: measures that would permit the timely detection of the diversion of nuclear material and the use of equipment for unauthorized purposes. Initially, there were bilateral safeguards applied by the exporters themselves. In the early 1960s, when the International Atomic Energy Agency (IAEA) had drawn up its first safeguards system, the United States began to transfer to that agency the responsibility for applying safeguards pursuant to these agreements for cooperation, and soon other suppliers, notably Canada and the United Kingdom, followed suit. In this way, these nations aimed at promoting the peaceful development of nuclear energy—and the role of their industries in that development—while ensuring that their international supplies would not be used for the proliferation of a military nuclear capability by the recipients.

These safeguards were first applied in 1960 to a small research reactor in Japan. In succeeding years, the safeguards system was successively extended to cover larger research reactors (thermal output of less than 100 MW), then small power plants, then power reactors of all sizes; in 1966, to reprocessing plants; and in 1968, to nuclear material in conversion and fabrication plants. Safeguards thus spread vertically, embracing all the major components of the commercial nuclear-fuel cycle.

The next phase came with the entry into force of the Treaty on the Non-Proliferation of Nuclear Weapons, or NPT, in 1970. The NPT divides the world among the nuclear powers, which commit themselves not to assist any other country in producing an explosive device; and all other parties to the treaty, which renounce such production and agree to place their entire nuclear activities under IAEA safeguards. The NPT thus marked a transition from the earlier safeguards system designed essentially to apply safeguards only to individual nuclear plants. The NPT envisaged what are now called "full-scope safeguards" that cover *all* nuclear activities of a country and are aimed at controlling the flow of nuclear material.

It is evident that the NPT embodies all facets of nuclear concerns of the 1980s: the division between nuclear have and have-nots, the wish to assure nonproliferation, the free acess to transfers of technology, and the right to assure fuel supply.

In order to implement the conditions of the treaty, forty-seven nations met under the auspices of the IAEA in 1970, for more than one year, to arrive at an international consensus on a model safeguards agreement, to be accepted by the non-nuclear-weapons states (NNWS). The technical objective of safeguards was defined as "timely detection of the diversion of significant quantities of nuclear material . . . and deterrence of such diversion by the risk of early detection."[8] Clearly the objective is deterrence. In other words, the value of safeguards is in their exposure of any diversion, and the continuing assurance they can give that no such diversion is taking place. This is the essence of the contribution of international safeguards to international security.

The basic bargain of the NPT was that non-nuclear-weapons states (NNWS) were assured access to nuclear technology and fuels in exchange for a no-weapons pledge and safeguards on all their nuclear activities. The application of safeguards has evolved since 1970, and has had an impact on international trade. Already by 1974, fourteen industrialized countries met in what became known as the Zangger Committee to elaborate upon the requirements of NPT Article III.2 concerning exports of nuclear equipment. The lists of equipment that were drawn up and were to trigger safeguards did not deny the need for export; they only sought to establish the criteria for nuclear exports. The lists served nonproliferation purposes because they sought to impose safeguards conditions on supplies that could be used for explosive purposes. The lists were formally adopted in June 1974, when after the Indian nuclear test of May 1974, the prohibition was extended to cover both nuclear weapons and nuclear explosive uses of supplies.[6] The lists represented the first major breakthrough in the effort to get the principal nuclear suppliers to accept the notion of establishing common conditions for the supply of nuclear equipment and materials.[9]

Export Controls

In 1975, Canada, West Germany, France, Japan, Great Britain, the United States, and the Soviet Union, representing the industrial nations responsible for most of the nuclear trade, met in London to work on an agreed set of rules tightening nuclear-export controls and procedures. This was a recognition by the major suppliers that it was in their interest to avoid the erosion of technical nonproliferation restraints through commercial nuclear competition. They finally agreed on the London Nuclear Suppliers' Guidelines, a common understanding on export policies of certain categories of nuclear equipment, material, and technology that was adopted in the summer of 1978 by Belgium, Canada, Czechoslovakia, France, Federal Republic of Germany, German Democratic Republic, Italy, Japan, the Netherlands, Poland, Sweden, Switzerland, the United Kingdom, the United States, and the Soviet Union. These guidelines, however, did not cover sales or transfers that are *not* dependent on technology acquired from one of the major suppliers, nor was there a common policy developed to exercise full-scope safeguards. In addition, only the Soviet Union, Poland, and the German Democratic Republic (East Germany) supported the principle that "trigger list" items should be exported only if *all* nuclear activities in the recipient NNWS are under IAEA safeguards.[10]

Until 1976 it was the view of the then-U.S. Atomic Energy Commission, and of the nuclear industry on both sides of the Atlantic and of the Pacific, that the separation of plutonium from spent reactor fuel, its recycling for use in conventional reactors and later for fuel in the expected breeder reactors, represented a logical move. The assumption was that plutonium would become a natural part of the nuclear-fuel cycle. India's explosion of a nuclear device in May 1974 is said to have shattered confidence in this approach, and has been described as one of the reasons for the concern that nuclear weapons would spread to less industrialized countries in the coming years.

Henry Kissinger in 1974 sensed the coming mood in the United States when he said to the United Nations that political and moral overtones might have to override economic considerations in the coming discussions of nuclear proliferation.[11] The statement preceded President Ford's comprehensive American program in October 1976 for action on nonproliferation both on the domestic and on the international levels.

Congress, too became increasingly aware of the problem. The Symington and Glenn amendments in 1977 cut economic and military assistance to NNWS that had acquired unsafeguarded enrichment and reprocessing technologies.

Both Presidents Ford and Carter called for a pause in plans to recover plutonium from U.S. spent fuel. President Carter in 1977 carried the move even further with an indefinite domestic deferral of reprocessing and of the commercial development of the breeder reactor. This trend culminated in the Nuclear Non-Proliferation Act (NNPA) passed in April 1978, about four years after the Indian detonation.

The NNPA is concerned with the assurance of adequate supplies of low-enriched uranium fuel, but imposes new and restrictive nonproliferation conditions on specific export licenses. It lays down the requirement that any country that imports nuclear material, equipment, or technology from the United States must have all of its nuclear activities under safeguards. Thus, the NNPA adopts the full-scope safeguards approach, a welcome feature of U.S. policy, which, if fully applied, could lead to the universalization of safeguards.

But the intent of the NNPA is to restrict the spread of reprocessing and enrichment facilities and to discourage commitment to plutonium fuels for recycling or breeder reactors. In a general attempt to impose this policy, the NNPA also demanded that countries trading with the United States under terms of existing bilateral agreements would have to renegotiate these agreements according to the newly imposed conditions.

In passing this piece of legislation, many believe that the United States attempted to use its monopoly in nuclear supply, above all its monopoly in enrichment services, as leverage to enforce its views on its nuclear trading partners, whether allies or not, although it was maintained that "any policy pursued to the point of severely shaking those alliances would be a failure in non-proliferation terms."[12]

From abroad, however, these efforts were seen as a unilateral action that legistlatively entrenched U.S. policy in a uniform set of restrictions. Whatever negotiating flexibility the United States had was limited to the exception procedures of the act. This prospect has tended to unify opposition to U.S. efforts, and Japan joined the states of the European Communities and some of the developing countries in opposition.

Perhaps one of the best overall assessments of the Nuclear Non-Proliferation Act came from a U.S. Senator who said that Congress had given its philosophy on nonproliferation, and that it was the task of the Administration to put it into effect to the best of its ability.

The International Fuel Cycle Evaluation

In the search for tighter controls over nuclear proliferation, the International Fuel Cycle Evaluation (INFCE) was launched by President Carter in November 1977 to study the technical and institutional problems of organizing the nuclear-fuel cycle in ways that provide energy while minimizing the risk of weapons proliferation. The basic premises were that nuclear power should be made widely available to meet the world's energy needs, that effective measures should be taken to minimize the danger of proliferation without jeopardizing energy supplies or the peaceful development of nuclear energy, and that special consideration should be given to the needs of developing countries.[13]

When INFCE started, there were wide differences of views on either side of the Atlantic and Pacific on the important questions of reprocessing, plutonium, spent-fuel storage, recycling, and fast breeders. In this context it can be noted that INFCE's first communiqué (Washington, December 1977) stressed that INFCE was to be a technical and analytical study and not a negotiation.

While the United States supported the idea of the once-through system as a "technical fix" that would prove proliferation-resistant, the Western Europeans and Japan wanted reprocessing because they could foresee scarcity of nuclear fuel and services. It seems that North American policymakers recognized during INFCE that for certain countries, reprocessing and the fast breeder would be indispensable. On the other hand, there was a realization that although the fast breeder might indeed be indispensable, the time scale for commercial reprocessing and for large-scale use of fast breeders was longer than originally foreseen.

INFCE has been useful in bringing countries to work jointly to try to restore the assurances of supply that have been undermined in the last few years, and to cooperate in dealing with the problems of nuclear proliferation. This two-year endeavor produced some 20,000 pages of reports and comprised some 200 international meetings. It is the most complete review of the *technical* problems of nuclear-fuel cycle since the Salzburg Conference on Nuclear Power and its Fuel Cycle in 1977. It is important to note that environmental and safety issues were also considered to determine whether specific fuel-cycle activities could be carried out in conformity with accepted standards, and it was found that these standards could be respected. It was also pointed out that by employing the present technology, radioactive wastes can be managed and disposed of with a high degree of safety and without undue risk to man or the environment.

The final conference of INFCE in February 1980 recognized that the time had come for political decisions on how to strengthen the nonproliferation regime and how to ensure security of supplies. The exercise confirmed the very limited scope of technical solutions and led back to the conclusion that proliferation is in the first instance a political problem, the solution of which lies in appropriate policies of consensus and cooperation, based on the goodwill and determination of all nations concerned to maintain peace.

In sum, INFCE has led to a greater understanding of proliferation risks, not only by participating governments, but also by the nuclear industry. It can be said that it supports the continued development of the breeder and some use of plutonium, the application of uniform controls and conditions for nuclear exports as well as the reliability of nuclear supply, and the reliance on institutional means for limiting nuclear proliferation. The prevailing outlook is optimistic for nuclear energy, and the viewpoint of the Carter Administration that nuclear power is an energy source of last resort will find little future support.

On the institutional side, interest in three tangible needs emerged from INFCE: (1) an international system for the storage of separated plutonium (this scheme was already foreseen in the original IAEA statute, and with strong support from Great Britain as well as several other nations, an IAEA study of the matter was started even before INFCE was launched); (2) arrangements for storing and managing unreprocessed spent fuel; and (3) a new set of ground rules to restore confidence in nuclear-supply assurances.

In considering the future, it seems unavoidable to recognize that a tighter linkage will be required between assurances of supply and assurances against nuclear proliferation. An improved nonproliferation regime should include a bouquet of measures:

1. Full international agreement on the measures required to strengthen nonproliferation arrangements;
2. Irrevocable guarantees of supply in return for irrevocable safeguards;
3. A condition that the sole nontechnical ground for interruption of supply would be the breach of a safeguards agreement;
4. Strict adherence by all suppliers to the nonproliferation conditions agreed upon, which should be the only limitation to purely commercial competition; and
5. A degree of restraint in the building of new sensitive facilities, with a preference for multinational solutions where feasible.

The closing conference of INFCE, held in February 1980, seemed to many to be an appropriate time to begin reconsidering nuclear-supply policies. A first step in this direction was taken in June 1980 when the IAEA Board of Governors established a Committee on Assurance of Supply (CAS). This committee, which is open to all member nations, must advise the Board of Governors on ways and means by which supplies of nuclear material, equipment, and technology, as well as fuel-cycle services, can be assured on a more predictable and long-term basis in accordance with mutually acceptable considerations of nonproliferation.

The second review of the NPT in August 1980 did not provide the opportunity for a consensus declaration by the parties to the treaty. One can only hope for the future of international nuclear trade that CAS (which includes countries not party to the NPT) will find it possible to build upon common ground. What is particularly needed is the same climate of concord and goodwill that had beome known as the "Spirit of Vienna" at the time of the first safeguards committee's meeting at the IAEA (June 1970 to March 1971), when many diverse interests had to be harmonized before a consensus could be found in the form of a model safeguards agreement that is now accepted by more than 111 nations.

Notes

1. A 1,000 MWe nuclear plant operating at 70 percent load factor will save 1.5 million tons of oil per year, or 30,000 barrels of oil per day.

2. Robert Stobaugh and Daniel Yergin, *Energy Future* (New York: Random House, 1979)

3. *Report of the President's Commission On the Accident at Three Mile Island* (Washington, D.C.: GPO, 1980), p. 8.

4. Communication from the EC Commission to the Council, "Energy Policy in the European Community: Perspectives and Achievements," COM(80) 397 final (Brussels: July 10, 1980), p. 13.

5. The declaration of the Tokyo Summit, June 22-23, 1979, is found in the *Department of State Bulletin* (August 1979), pp. 8-9. For the declaration of the Venice Summit, June 22-23, 1980, see *Department of State Bulletin* (August 1980), pp. 8-11.

6. The limitations of Western Siberian oil supplies is also evident from the USSR 1981 Annual Plan and the draft Five Year Plan 1981-1985. The planners have called for strict conservation measures and the conversion of oil-fired electric stations to coal.

7. Dwight J. Porter, Director, American Nuclear Energy Council, Testimony before the House Interior Subcommittee on Energy and the Environment, U.S. Congress (Washington, D.C.: July 26, 1979). For further reference see Senator James A. McClure, "U.S. Nuclear Power Isolationism, a Case Study in the Abdication of World Leadership," address before the Commonwealth Club of California, San Francisco, February 1979.

8. INFCIRC/153: The Structure and Content of Agreements between the Agency and States required in connection with the Treaty on the Non-Proliferation of Nuclear Weapons. Information Circular issued by the International Atomic Energy Agency (Vienna: 1971).

9. The Zangger Committee trigger lists are reproduced as Appendix 1 in Ben Saunders, *Safeguards Against Nuclear Proliferation*, SIPRI Monograph (Cambridge, Mass.: MIT Press, 1975). They were distributed to IAEA member states in INFCIRC/209 (Vienna: IAEA, 1974).

10. The London Suppliers' Group is discussed in Joseph A. Yager, *International Cooperation in Nuclear Energy* (Washington, D.C.: Brookings

Institution, 1981), pp. 34-36. The London Guidelines are reproduced in Appendix D of *Nuclear Power and Nuclear Weapons Proliferation*, vol. 2 (Washington, D.C.: Atlantic Council of the U.S., 1978), pp. 63-74.

11. A/PV.2238, September 23, 1977, XXIX Session of the General Assembly of the United Nations.

12. Joseph S. Nye, "Non-Proliferation: A Long-Term Strategy," *Foreign Affairs* (April 1978), p. 619.

13. Final communiqué of the organizing conference of the International Nuclear Fuel Cycle Evaluation, December 12, 1977, and list of INFCE participants in *INFCE Summary Volume* (Vienna: IAEA, 1980).

5 World Coal Prospects

William F. Martin

Coal has emerged as the most important alternative to our dependence on oil. It cannot carry the burden alone. Energy conservation and nuclear power are also important. But studies made by the International Energy Agency (IEA) and the World Coal Study of the Massachusetts Institute of Technology (MIT) suggest that coal use could triple over the next two decades, meeting up to two-thirds of the world's additional energy requirements of that period. Coal's share of total primary energy could increase from 20 percent today to 35 percent in the year 2000. At the same time oil's share could decline from 50 percent to between 25 and 30 percent of total primary energy.

To realize the full potential of coal, a major steam-coal trade must be developed. As much as 700 million tons might be traded by the year 2000. In terms of energy equivalents, this is about the level of Saudi Arabian oil exports today.

Quite importantly, the major producer of coal in this scenario would have to be the United States. By the year 2000 the United States could produce 2 billion tons of coal, of which 300 million tons could be provided for export to Europe and Japan.

The World Energy Outlook

As a background to a more specific discussion of coal, it is useful to examine the longer-term world energy outlook. A plausible scenario of the world energy future in the years 1990 and 2000 will be presented. The scenario is based on the assumptions of modest economic growth of approximately 3 percent by the end of the century, and energy demand growth of about half the rate of economic growth. This would lead to about a 30 percent increase in overall energy demand over the course of the next two decades. What then are the significant supply sources that will meet this level of energy demand?

Nuclear energy has to be a major contributer. In order to reduce overall dependence on oil between now and the end of the century, nuclear energy must be vigorously pursued. The IEA estimates that if 15 Gigawatts GWe were to be ordered annually over the next decade, nuclear energy by the year 2000 could increase almost fivefold and in doing so contribute about 15 percent of total energy needs of industrialized countries.

But serious problems remain—especially public acceptance. Though nuclear energy grew at a rate of about 25 percent during the late 1960s and early to mid-1970s, growth in 1979 and 1980 has only been 2 percent per year. Yet there are some positive signs of recovery of the nuclear option in Europe and the government in the United States has indicated a strong preference for nuclear energy as an indispensable contribution to overall energy supply.

Second, *oil and gas* production levels within the IEA would have to remain stable between now and the year 2000. Given the reserve position and higher prices, this objective is achievable. Part of the effort required to ensure this objective will be to encourage exploration for new discoveries; but in addition, and perhaps more importantly, there is great potential to recover more of the already discovered oil through better enhanced-recovery techniques.

Third, *new and renewable energy sources* would have to contribute about 5 percent of total supply. Though these sources could become increasingly important after the year 2000, their contribution before then is modest because of long lead times in their development and commercialization.

Last, to successfully complete the scenario, it will be necessary to double *coal production and use* by the early 1990s and triple it by the year 2000. The implication of this scenario is that the share of oil consumption in total energy consumption—today about 50 percent—would fall to 40 percent in 1990 and about 25 to 30 percent at the end of the century. Net oil imports to IEA countries would fall from 22 mbd today to 20 mbd by 1990 to 15 mbd by the year 2000. Oil would be increasingly reserved for transportation and nonenergy uses and electricity would have to grow from one-third of total primary energy today to over 40 percent by the end of the century, requiring an electricity growth of about 3 percent per year.

The Role of Coal

Against this background, coal must be viewed in a new light. Given some uncertainty over nuclear-power expansion; given limited sites for hydroelectric-power expansion; and given a probable leveling of oil and gas production during the 1980s, coal has the potential to provide the foundation for future world economic activity.

There are numerous reasons why coal can become the most significant substitute fuel for oil. First, reserves are abundant and widely dispersed. The world has recoverable reserves of 640 billion tons—enough for 250 years of consumption at present levels. Reserves are well-distributed between industrialized countries of the OECD, the Soviet Union, Eastern

Europe, China, and numerous developing countries. Production of coal should therefore not be a serious constraint. The major challenge will be to find markets for expanded production and to overcome problems of inadequate infrastructural development in moving coal from producer to user. Important environmental and safety issues must also be resolved.

The Emerging Coal Future

It is instructive to review the last two years to see how coal demand, production, and trade infrastructure have developed. Following are some conclusions about the status of coal development in IEA countries.

Coal Demand

Compared with the price of oil, coal is now very economically attractive for electricity generation. A recent report of the U.S. Task Force on Coal Exports indicated that in the present circumstances it could be cheaper to build a new coal-fired power plant for baseload than to continue to operate an existing oil-fired plant. Coal delivered to Europe is much less expensive than imported oil, even after including higher transportation and treatment costs. The Japanese are able to import and burn coal in an environmentally acceptable manner at half the cost of imported oil.

Policies in IEA-member countries have indicated strong growth potential over the next decade and coal is expected to increase by 60 percent in electricity generation. The IEA Coal Industry Advisory Board has concluded that implicit in doubling coal use by 1990 and tripling it by the year 2000 is a requirement for coal-fired generating capacity within the OECD countries to increase from the current level of 350 GWe to about 1,100 GWe by the end of the century. To achieve this level countries with little coal-generating capacity to date will have to increase their commitments. This is especially true for Italy, Japan, and the Netherlands.

Some troubling questions still remain about future penetration of coal in electricity generation. There are uncertainties about the rate of growth of electricity demand; the financial position of utilities, and the contribution of nuclear energy. All of these factors will influence the ultimate contribution of coal.

Economics is the key to inducing some industries to substitute coal for oil. The cement industry, for example, has experienced a dramatic shift from oil to coal. In some cases the pay-back period is only two years—a competitively attractive option given the rising costs of oil. By 1985 it is estimated that close to 90 percent of the energy needs of the cement industry in Europe will be met with coal, compared with only 15 percent in 1973. In Japan this industrial sector will be converted almost entirely to coal by the end of 1981.

The possibilities of expanded coal use in other industrial sectors are less well known. IEA countries have indicated that the use of solid fuels is likely to increase by more than 70 percent from 1981 to 1990. However, uncertainties exist about the costs of conversion, pay-back periods, and ability of industry to finance such conversions particularly in current economic circumstances. Coal can be burned in most applications in ways that are environmentally acceptable at reasonable costs. But in some cases environmental procedures are complex and cumbersome, and the very complexity of the process can act as a disincentive to switching from oil to coal.

The potential of synthetic fuels derived from coal must still be regarded with some caution because of technological and economic uncertainties. However, it is imperative that we continue to test, refine, and develop these technologies to ensure that some of them are in a position to supplement world energy supplies by the 1990s.

Several other promising developments loom on the horizon. Improved interaction between energy and regional-development planners could foster the development of industrial parks where the energy requirements of many smaller firms could be supplied by central coal plants, giving them full advantage of the scale economies and higher efficiencies offered by cogeneration.

Coal Production

Most studies indicate that coal production is unlikely to constrain coal development given the large and relatively well-dispersed resource base.

Producing countries can meet climbing demand in consuming nations if there are assurances of future markets and long-term contracts to "bankroll" required investments. If these are forthcoming than production and infrastructure are likely to follow. By 1990 Canada could double production; the United States has the potential to increase coal production by some 50 percent; and Australian production could climb two-and-one-half times.

World Coal Trade and Infrastructure

Since demand and production are rising rapidly, it is understandable that a number of problems of inadequate or incomplete infrastructure have occurred. In some cases short-term solutions were successfully applied. In others, constraints prohibited effective functioning of the coal network. For example:

In 1980 U.S. exports of coal were up by 40 percent. Given the short-term rise in domestic and overseas demand, the performance by industry in using existing infrastructure to meet the surge in coal production and use has been remarkable. Yet problems have occurred, the most notable of which has been congestion of port facilities resulting in higher costs for buyers.

In Canada, exports in 1980 were up by some 8 percent but financial difficulties in developing mines and transportation systems (both railroads and ports) have hindered industry.

In Australia, exports increased some 7 percent in 1980 but further development is slowed by continued uncertainty about potential export demand and by inadequate railroad systems. There have also been recent strikes in Newcastle by dockworkers, which have lead to port congestion.

In Europe, port facilities are generally adequate to handle present coal imports, and progress is being made to expand facilities. However, some bottlenecks are beginning to build up. Denmark, Germany, the Netherlands, Spain, and France have all made progress in this area. Italy, Portugal, and Turkey have lagged behind, largely because of financial problems.

Japan has a comprehensively planned program to increase coal imports. Financial assistance has been granted by the government to ensure its successful implementation. A two-million-ton-per-year coal center is currently under construction and two additional centers are to be completed by 1985.

On the whole, these developments are impressive and suggest that progress to increase coal use on a substantial scale is under way. Some of the difficulties experienced in 1980 reflect normal growing pains of a much more rapid shift to coal than had been anticipated, and actions are being taken to overcome such difficulties; other difficulties reflect more fundamental areas of uncertainty and require further consideration.

Environmental Issues

A longer-term constraint on the coal industry is how to greatly expand coal production, trade, and use in an environmentally acceptable manner. This problem is particularly acute for surface mines, which must be relied on for a very substantial proportion of increased global coal production.

Surface mining imposes considerable demands on the local environment, so that rigid control over material movement, restoration of the topsoil, and landscaping of the mine area must be maintained. An example of the successful handling of these problems is the recently developed mine in

Wyoming known as Black Thunder, which at any one time will disrupt only 250 hectares for the total mining operation. Its planned annual physical production is over 20 million short tons, or the equivalent of 200,000 barrels of oil per day. Such production on relatively little land enables the producers to reinvest heavy sums in land reclamation and to carry out such reclamation rapidly.

Perhaps the world's most impressive coal development with regard to community and environmental concerns exists in the Rhine area of Germany. During the last three decades, more than fifty communities west of Cologne containing more than 24,000 inhabitants have been successfully relocated to make room for mines. The major coal developer in this region, Rheinische Braunkohlenwerke, has never had a major open-pit coal-mine project rejected and generally has had a very successful record in relocating populations to their satisfaction.

In the case of deep mining, *safety* remains an important issue. In Britain today, it is estimated that deep-coal mining entails a loss of life at a rate of one person for every two million tons of coal mined, while the U.S. experience is roughly comparable. Plainly, every effort must be made to reduce these figures.

Deep coal mining also has its own environmental problems. The amount of spoil per ton of coal produced varies greatly from mine to mine, but the quantity might be 30 percent of the coal produced. Existing techniques and new technology under development, including the use of new fluidized combustion technology to burn waste at the mine mouth to produce useful power, can make these problems manageable.

Next there is the problem of *solid wastes* at the combustion site, wastes which may be quite unfamiliar to industries replacing oil with coal. A 1,000-MW power plant may have to dispose of 320,000 tons per year of ash. There are useful means for its disposal however. Fly ash can be used to make cement, and is useful in road construction. Another solution that has been adopted in Japan, and could find application elsewhere, is to use the ash for creating new industrial sites in shallow water offshore.

More serious is the problem of *pollution,* especially from sulphur oxides. This problem is partly overcome by the fact that most of the incremental world coal production foreseen is of a low-sulfur quality, thus the proportion of high- and medium-sulphur coal is expected to sharply decline. The technology for keeping sulphur-oxide emissions to environmentally acceptable levels is available and in use today. The problem is primarily one of cost.

At this stage concerns about atmospheric carbon dioxide buildup (a consequence of all fossil-fuel burning, not only of coal use) are not sufficently documented to justify delays in expanding world use of coal, although undoubtedly this issue will have to be kept under close surveillance.

In conclusion, a tripling of coal production and use does not pose insurmountable environmental problems. With the application of technology already available, coal can be mined, moved, and used in most areas in complete conformity with high standards of safety and environmental protection. And the increase in cost this involves is not unacceptable, especially when compared to the present price of oil.

World Steam-Coal Trade

One of the greatest challenges to increased coal use worldwide is the development of a major steam-coal trade. Tripling world steam-coal use within industrialized countries would require a tenfold increase in steam-coal trade to a level of a least 700 million tons per year.

What will be the direction and magnitude of the increased trade in coal? The World Coal Study estimates that by the end of the century U.S. exports could rise to some 300 million tons of coal equivalent (mtce); other major exporters would be Australia and South Africa, possibly trading at a level of 250 and 150 mtce per year respectively in the year 2000. Other exporters may include Canada or even China. Major importers would include Western Europe (500 million tons per year) and Japan (200 million tons per year).

How might this scenario be realized? First and foremost, the prospect must be faced in energy-deficient countries that oil imports will simply not be available to meet incremental demand. Consequently, these countries will have to depend on coal to meet a growing proportion of their energy needs. Likewise, coal-rich countries will have to accelerate present development strategies so that resources are exploited to fulfill both domestic and possibly sizable overseas demands.

To these ends, ministers from IEA countries in May 1979 agreed to Principles for IEA Action on Coal and a system to monitor progress in implementation of these principles. At the heart of the principles is the objective of expanding world coal trade by providing a governmental umbrella to facilitate commercial negotiations between consuming and producing countries. In the principles a careful balance was sought between the interests of consumers and producers.

On the one hand, potential coal-importing countries agreed to adopt strong coal utilization strategies; to impose no new restrictions on imports; to implement measures to support domestic production; to ensure that no new measures restrict coal consumption; and to encourage development of the required trading infrastructure.

On the other hand, the agreement encouraged potential coal exporters to produce coal for export; to establish an investment climate conducive to coal development; to promote development of the required infrastructure; to impose no new restrictions on coal exports; to operate existing coal-export-control systems so as not to limit the growth in overall coal consumption; and to refrain from interference in long-term contracts.

These are important steps. They represent the first coherent formulation of joint multilateral policies to promote coal. But major problems have still to be overcome before the necessary balance of interests can be attained.

Information. Better *information* about expectations of producers and consumers is necessary. For example, in Australia, where mines are currently producing at capacity, three to seven years are needed from the time a long-term contract is signed to first deliveries. Several consuming nations that have forecast significant increases in coal imports in 1985 and 1990 have not obtained coverage for all of the anticipated volume. This problem can be partly overcome by more widespread international communication regarding coal plans, so that producers and consumers can be aware of requirements early enough so that action can be taken in a coordinated way to avoid potential bottlenecks in the coal chain.

Infrastructure. Most IEA countries do not currently have adequate capacity to transport, export or import, handle and store significantly greater amounts of coal. Some progress is being made. Denmark has developed port facilities accessible to large bulk carriers. Spain is making progress in converting two of its ports to handle increased coal imports. Ireland is building a deep-water coal harbor. In most producing countries there is a lack of adequate rail facilities to bring coal supplies to port. Of particular importance in the United States is the need to make a decision soon on dredging access channels to ports. New port capacity and other means for increasing loading capacity must also be decided upon promptly. Government and industry are very much aware of these problems and are working to eliminate them. I understand plans are now under way in the United States to expand port capacity by some 23 million tons by 1983. This is encouraging, but if we are to achieve a level of U.S. coal exports of 300 million tons by 2000, more must be done.

Investment. A stable climate for international investment is vital. The World Coal Study estimated that the total capital investment to produce, move, and use coal would come to about $1,000 billion. A major proportion of investment capital will necessarily come from private sources, and investors' decisions are bound to be influenced by their perception of the economic and regulatory climate during the life of the investment. As far as possible, governments should avoid excessive revisions in legal, administrative, or environmental regulations that would disproportionately increase costs and reduce confidence.

Prices. Since it takes from seven to ten years to develop large-scale coal-using facilities, there is notable and understandable concern on the part of consumers as to the long-term price trend. Spot prices have risen over 70 percent in the last two years, although coal is still highly competitive with oil. Some potential buyers are uncertain about whether markets will operate competitively and, in particular, about whether governments of coal-exporting countries might impose taxes on coal for export with a view to moving its price closer to the price of oil. Over the longer term, as world trade develops, and if short-term bottlenecks can be eliminated, most observers feel that the price of coal—determined by market conditions of supply and demand—should remain considerably lower than the price of oil.

Protection of Domestic Industry. The IEA has been saying for some time that countries should not block the import of coal, particularly to displace oil that could be used beyond the capacity of domestic production. This is becoming less of a potential problem. Germany has recently expanded its import quotas, and there are no administrative controls on coal imports into the United Kingdom.

The IEA has been working to overcome some of these problems. In addition to country assessments which are conducted annually, the IEA has established a Coal Industry Advisory Board composed of senior representatives of industry involved with coal production, use, and transport from major industrialized countries. Their major conclusion is that despite a new encouraging attitude toward coal, considerable uncertainty exists about its future role. Unless commitments are backed up quickly by stronger measures, particularly on the use side, doubling of coal by 1990 is unlikely. However, if stronger measures can be taken by government and industry then the goal of tripling coal use and production is attainable, especially if coal electrical-generating capacity can triple over this period.

Conclusion

Let us conclude with some lessons learned over the last few years about coal.

First, coal, which once fueled the industrial revolution, is gathering momentum to provide the single-most-important energy contribution to the future of modern societies. Given uncertainties about the future role of nuclear power and the long lead times for commercialization of new and renewable energy sources, the conclusion is inescapable that coal must play a vital role in our energy future.

Second, the challenge of the next twenty years is to put in place enough coal to meet world energy demands equivalent to 40 million barrels per day of oil. The potential exists, but there are serious obstacles to be overcome.

While the last seven years have seen a great deal of progress in arresting the coal decline, we have only just begun to make the structural adaptations necessary to realize coal's full potential. Until now, expanding coal production and use has largely meant stretching existing mines, infrastructure, and user equipment to full capacity—and sometimes beyond. Very soon we will reach the limits and new capacity must be put in place, just about everywhere. This will be costly and time-consuming but the rewards will be great, and the penalties for not taking action almost unthinkable.

Third, political support at the highest level is behind the coal option. But general commitments need to be supported by specific measures. In particular it is essential that governments do their share by pursuing consistent coal policies over time, and by not confronting industry with basic changes in attitudes as day-to-day politics sometimes dictate. Industry, on the other hand, has the responsibility to ensure that the job is completed in a timely and efficient manner. There are some important challenges to be overcome in linking the coal chain, yet they are not insurmountable. The economics are favorable; the technology is available; and major environmental problems in production and use are solvable. However, coal development of the type envisaged cannot be based on an extrapolation of the coal economy of the past. It is a new challenge; it will take a new approach.

References

Coal, Bridge to the Future, Report of the World Coal Study, (Cambridge, Mass.: Ballinger Publishing Company, 1980).

Steam Coal Prospects to 2000, International Energy Agency (Paris: OECD, 1978).

Report of the IEA Coal Industry Advisory Board, International Energy Agency (Paris: OECD, 1980).

6 Alternative-Energy Technologies: The International Initiatives

Niels de Terra

New technologies that will enable us to use energy more efficiently and provide us with energy from sources other than imported oil have received ever-increasing attention since the 1973-1974 oil crisis. Alternative-energy technologies cover a very wide range, providing energy sources not based on conventional oil, or enabling us to use energy more efficiently. Examples of such technologies in the production of energy include wind power, geothermal power, and the extraction of oil from shales and tar sands. At the conversion stage, when energy is transformed from one form to another, the technologies include coal liquefaction and gasification, alcohol fuels from biomass, and advanced converter and breeder reactors. In the end-use sector, alternative-energy technologies include solar heating and cooling, new automotive engines and fuels, and heat pumps that could provide space heating from electricity far more efficiently than conventional devices.

Judging from past experience, it can take twenty years for a technology to move from the stage of research and development (R&D) to the point where it can be said to be in general commercial use. For the simpler technologies, such as solar heating and conventional heat pumps, the time span is less, and for the most complex and advanced technologies, such as nuclear fusion, the time span will be much greater.

New energy technologies are today being developed by governments and by the private sector. Typically, those technologies involving a high technical and economic uncertainty and long lead times are developed with government money, whereas those with lower technical and economic risks and shorter development lead times are funded by private industry. Solar electric power is an example of the former; improved internal-combustion engines an example of the latter.

The need to reduce the use of oil has now become axiomatic in energy policy in nearly every country that imports part or all of its energy requirements in the form of oil. But there is a bewildering number of ways to use less oil and to each one belongs a set of advantages and disadvantges in the degrees to which they affect economic growth, environment, lifestyles, and political sensibilities.

The opinions expressed in this chapter are those of the author, and do not necessarily reflect the views of the International Energy Agency.

Planning for the Twenty-first Century

The further ahead one looks into the future, the more important the role of alternative-energy technologies becomes. For the near term, there is comparatively little that alternative technologies can contribute to reducing oil use. A family that bought a car last year may keep it another five years instead of replacing it with a new, more efficient, and more expensive car just on the market. A public utility that built a generating plant five years ago will want to keep that station running another fifteen years in order to amortize its investment rather than replace it with a plant that is more efficient, but more expensive as well. There are, of course, ways in which governments can stimulate changes in consumers' and investors' perceptions, but it is generally recognized that new technologies will not begin contributing in a significant way to reducing our dependence on imported oil until close to the end of the century. This is partly because they have not yet been developed to the point where they are reliable and economical, and partly because it takes time for industrial societies to replace their stock of energy-consuming equipment.

The choice of which alternative-energy technologies will be developed is usually made using all the normal political processes by which national priorities are decided. Such decisions evolve over time, and rarely involve sudden changes of direction. For example, in 1974 the U.S. government spent 75 percent of its energy R&D budget on nuclear technologies (conventional reactors, breeders, and fusion), while its 1981 budget allocated 44 percent to these technologies.[1] The change is explained by a steady growth in funding for energy conservation and renewable sources of energy.

However, different technologies will be needed by different countries, depending on their climate, endowments of energy and nonenergy resources, technological capabilities, and economic strength. Solar heating will be of less interest to a country that sees little of the sun, as breeder reactors will not for a long time figure in the energy futures of countries with less-developed technological infrastructures.

Prior to 1973 most of the Western industrial countries regarded energy research and development as being synonymous with nuclear R&D, and there were very few significant nonnuclear R&D programs under way. With cheap oil and gas it seemed that conventional nuclear reactors and advanced reactors (breeders, high-temperature reactors, and fusion) were the only areas requiring expenditures of public money. It was thought that private industry would take care of nearly all other needed improvements in energy-producing or conserving technologies under normal operation of market forces, that is, the probability of economic return would justify private investment in R&D. A major change in government-supported energy R&D began to appear by 1976 as a result of the 1973-1974 oil crisis and price increases. Whereas in 1974 the governments of the countries belonging to the International Energy Agency (IEA)[2] were spending $1,878 million on energy R&D, this had increased to $4,019 million (in 1974 dollars) by 1977, and to $4,529 million by 1979. By far the largest part of the increase was for nonnuclear RD&D: coal, conservation, and renewable energy sources. In current prices, the 1981 outlays of IEA governments on energy research, development, and demonstration (RD&D) will exceed $8 billion.

Three major questions for energy and economic policymakers must then be asked: One, is $8 billion enough, or should it be $18 billion; Two, is the money being spent on the right technologies; Three, what are the right technologies? In a world of increasingly stringent national budgets, reduced economic growth, competing priorities among national security, social welfare, industrial modernization, and new energy systems, how should governments invest their taxpayers' money to ensure that, a generation hence, those technologies will be available that are really needed at that time?

The size of national-energy R&D programs varies as greatly as the different emphasis given to alternative-energy technologies. The United States in 1980 budgeted $4.2 billion for energy RD&D with 6 percent going for conservation technologies, 17 percent for fossil fuels, 44 percent for nuclear energy, 17 percent for renewable energy sources, and 15 percent for supporting technologies. Japan, with the second-largest program in the IEA, budgeted $1.3 billion for energy RD&D, with a much higher proportion going to nuclear power (74 percent) because of the fact that Japan has virtually no domestic energy resources of its own and sees nuclear power as the main way to reduce its dependence on imported oil. The amounts other countries devote to developing alternative-energy technologies vary with their size and gross national products. Ireland spends approximately $6 million per year, Belgium about $125 million annually, and the United Kingdom close to $430 million on energy R&D.

It is difficult enough within one country to obtain a consensus on national priorities for energy technologies. Each technology has its advocates who advance their cause to the detriment of others: renewable sources are stressed because of their supposedly benign environmental effects, but criticized because of their presumed high costs and limited applications; nuclear power is advocated as the most economical alternative to oil-fired power generation, and opposed as being unsafe; coal conversion technologies are advanced as the best way to provide needed alternatives supplies of liquid fuels, but questioned on environmental grounds. If one looks at the Western industrial countries, those that have been able to reach a policy consensus on long-term energy technology priorities are the exception rather than the rule. As often as not, the de facto priorities prevailing at any one time reflect a political and economic bargain struck among interest groups.

Against this background of varying national perceptions of long-term energy-technology needs, with no common international perception among the industrial countries, the IEA set out early in 1976 to develop a long-term strategy for energy RD&D for its member countries. The original call for such a strategy came from part of the agency's charter, the Long-Term Cooperation Program, adopted by its members in January 1976.

The elaboration of an RD&D strategy fit with a new approach to energy problems taken by IEA countries since 1974. This was based on the premise that if each of the industrial countries pursued national solutions to the global energy problem derived solely from national perceptions of self-interest, the results would only be suboptimal, and might well be self-defeating. But if national solutions were influenced and guided by common perceptions of the problem and if these perceptions resulted in common objectives, undesirable effects might be avoided or at least minimized, and individual countries might be better off for it. One example of this approach is the IEA's emergency oil-sharing system. This is aimed at avoiding a repetition of what happened during the 1973-1974 oil crisis, when countries scrambled to ensure their own supplies of oil in a market that had been made very tight as a result of the oil embargo. This led to a situation in which countries were bidding up the prices of scarce oil supplies, and thus forcing each other to pay more for the oil than they would have had to pay if a way had existed to share oil among users on a equitable basis. The IEA's oil-sharing system, created since 1974, is intended to prevent a recurrence through an agreed system of conserving existing supplies and sharing new ones so that no one country suffers more or less than other IEA members.

A similar approach was conceived for energy-technology development. In this case, the potential benefit to member countries lay in a more rational and efficient use of resources through the creation and application of a strategic plan for the technological transition away from oil. The task was to examine the long-term (up to the year 2020) needs of the IEA group of countries for energy technologies under various likely assumptions as to what the dominant political, economic, and social trends influencing societies' choices of energy technologies might be. Through such analyses, projections and other information would be obtained that would provide answers in the form of a strategy to such questions as:

Which new energy production, conversion, and end-use technologies will the IEA countries as a group need over the next several decades, and how much of a contribution to future energy balances can these technologies be expected to make under different assumptions as to what the future might look like?

How should the national energy RD&D programs of the IEA be structured, in terms of level of effort and internal balance, to maximize the

IEA's chances of having the technologies it will need in sufficient quantity when they are needed?

In a nutshell, the IEA countries were implying that if there were technologies that could significantly reduce their oil-import dependence more than others, they could structure their R&D programs to emphasize such development.

The Development of a Strategy for Energy RD&D: The Analytical Phase

It is now common practice in government and business to use computer-operated models to study the behavior of very complex systems, such as economies and national energy systems. Thus, early in 1976 a system-analysis project was started that eventually came to involve close to 100 man-years of effort.[3] This work took place at two centers, one located at the Brookhaven National Laboratory in Upton, N.Y., and the other at the Nuclear Research Center (KFA) in Juelich, Germany. Fifteen IEA countries and the Commission of the European Communities assigned experts in energy-systems analysis to one of the two centers for the duration of the project, which lasted until the beginning of 1980. The teams contained approximately fifteen people each, including management and computer-support staff.

Building on previous modeling work undertaken at the two laboratories, a computer model was developed for the analysis called MARKAL. The model simulates the functioning of the entire energy system of a single country as illustrated in figure 6-1. It contains not only the detailed characteristics of a country's energy system, but also best estimates of the technical and economic characteristics of all existing and new energy technologies expected to become available over the next forty years. The model was used to analyze the behavior of national energy systems under varying assumptions. For example, it might be run in such a way as to simulate a political environment in which reducing oil imports was the most important single objective, and one for which society would be prepared to sacrifice some other goals. The model would indicate how many coal liquefaction plants, electric cars, solar water-heaters, and so on would be built or installed over the forty-year period under such a policy. Or it might be used to study the effects of a government policy aimed at making maximum use of renewable-energy technologies such as solar, wind, or geothermal.

Assumptions were made as to long-term economic growth rates, energy prices, energy demand for each demand subsector, the capital and operating costs, as well as technical parameters of all existing and prospective new

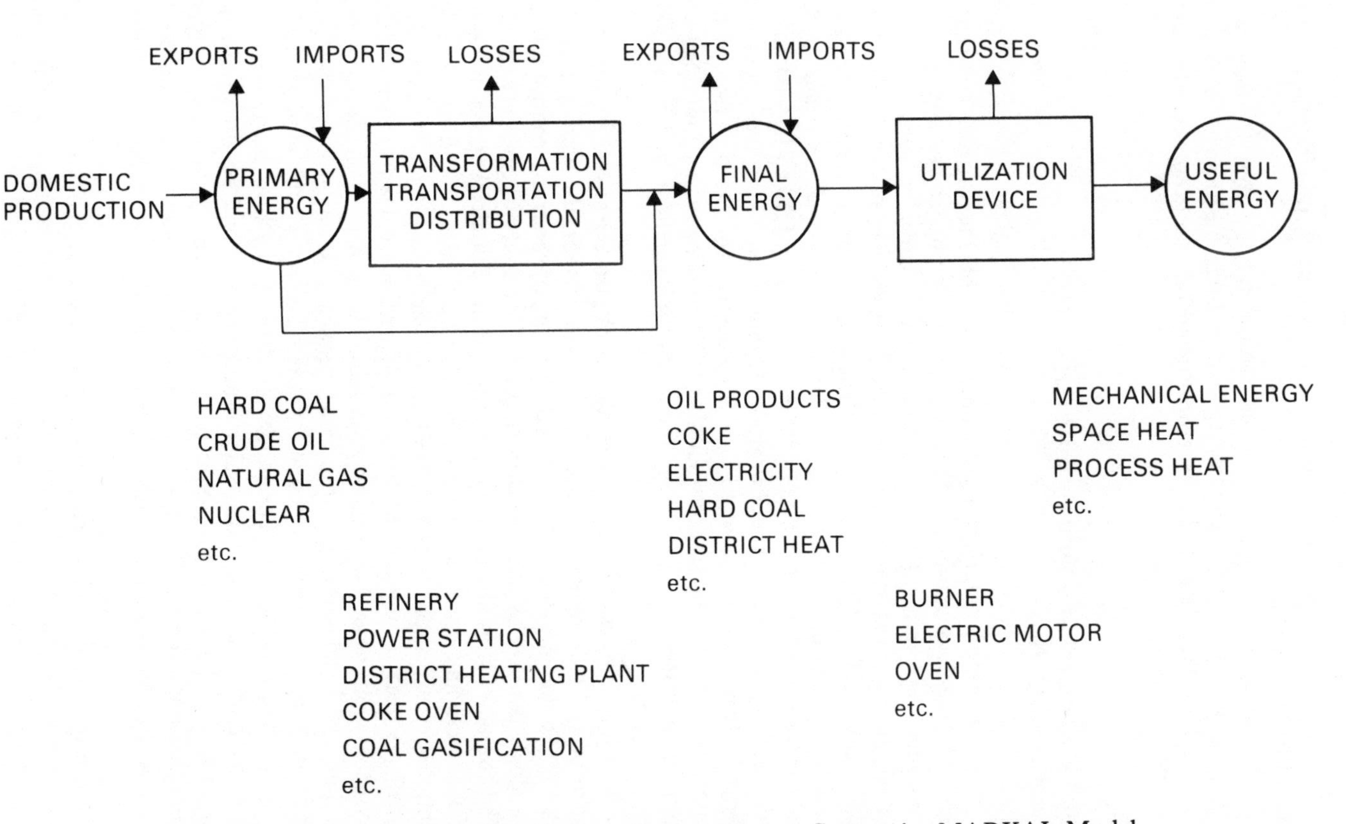

Figure 6-1. IEA Energy RD&D Strategy-Project Energy System in MARKAL Model

energy technologies. The MARKAL model was used to analyze a number of scenarios that were selected because they each represented either a central emphasis in energy policy (the objective of minimizing the cost of the nation's energy system, or the objective of keeping oil imports to a minimum), or because they represented a possible future that might be constrained in a particular direction for socio-political or technical reasons (such as limitation of the growth of nuclear power or of the use of fossil fuels). A policy placing maximum emphasis on renewable energy sources was also analyzed.

In addition to studying different assumptions about how future energy systems might be shaped by different policies, studies were also made of the effects of accelerating the introduction of new technologies on oil imports and energy-system costs. Such cases were intended to simulate the effects of major increases in energy-RD&D spending and of greater efforts to commercialize alternative technologies. The effect of such efforts was assumed to be that alternative technologies would become available sooner and in greater quantity than they would if current levels of effort were continued. For example, in a base case it might be assumed that coal liquefaction plants could not be built to start displacing conventional oil until 1995. In an accelerated case it might be assumed that with a larger effort starting now, such plants could be producing by 1990. Although sixteen scenarios were examined altogether, the principal scenarios were the following:

Minimum energy-system cost: A business-as-usual case in which it was assumed that the main factor influencing the introduction of new energy technologies into the energy system was a desire to keep the costs of the entire energy system to a minimum. It was also assumed that today's pace of technology development continued.

Limited nuclear power: In this case a future was studied in which it was assumed that the growth rate of nuclear power would be reduced to a level below that which might result if nuclear technologies were deployed solely on economic criteria. In fact, if the slowdowns in national nuclear-power programs being experienced today cannot be reversed, this scenario may be on its way to being realized.

Limited fossil fuels: These cases simulated a future in which the use of technologies using fossil fuels would be constrained by a problem such as might exist if increased levels of carbon dioxide were proven to be harming the environment, and limits placed on carbon dioxide emissions; or if some other environmental problem were discovered.

Accelerated security: A case assuming that the objective of increasing our security, that is, minimizing oil imports, takes precedence over minimizing energy-system costs. The reduction in oil imports is achieved in part by assuming that new technologies that displace imported oil are helped in the marketplace by measures that would assure their commercialization even if the cost of energy they produce was higher than the equivalent price based on prevailing world oil prices at the time they are introduced. It was also assumed in this case that technologies are being developed and commercialized at an accelerated pace; that RD&D efforts are substantially increased over present levels.

Maximize renewable energy sources: This scenario was designed to estimate the upper limits of the deployment of renewable-energy technologies, and it assumed that policy decisions would be implemented that would push the use of renewable energy regardless of cost.

Results of the Systems Analysis

A very considerable amount of information was obtained on what future energy systems might look like under the various scenarios studied. The essential summary data have been published in the already cited document, and more detailed results will be published by the two laboratories during the course of 1981. Results obtained through linear programming modeling and analysis must be used with care for various reasons. They are highly dependent on the accuracy of the input data, and although great care was taken with the preparation of the technical and economic characterizations of the new technologies, there are evident risks in projecting the costs of technologies decades away. Nevertheless, for the analysis, it is relative costs that are more important than absolute values. Furthermore, the analyses only show how systems might behave if one objective is pursued to the exclusion of all others, for example minimization of energy-system costs. In the real world the actual path is a permanent compromise between competing objectives. The value of such systems analyses is that we obtain useful insights into *how* particular policies affect energy systems in terms of technologies used, oil imported, capital invested, and so on.

The major conclusions of the MARKAL systems analysis are as follows:

1. *Oil imports decline in all scenarios during the 1980-2020 period.* However, during the first twenty years the decline is caused mainly by conservation and the expanded use of existing technologies. It is during the second twenty years that new liquid-fuel technologies begin to make a significant contribution. Oil imports under two scenarios are shown in figure 6-2, from which it will be seen that the reduction in imports during the remainder of this decade under the minimum-energy-system-cost scenario is quite unsatisfactory, but that under the accelerated-security scenario a very significant oil-import reduction seems possible. The difference between the two scenarios by the year 2000 amounts to over 7 million barrels

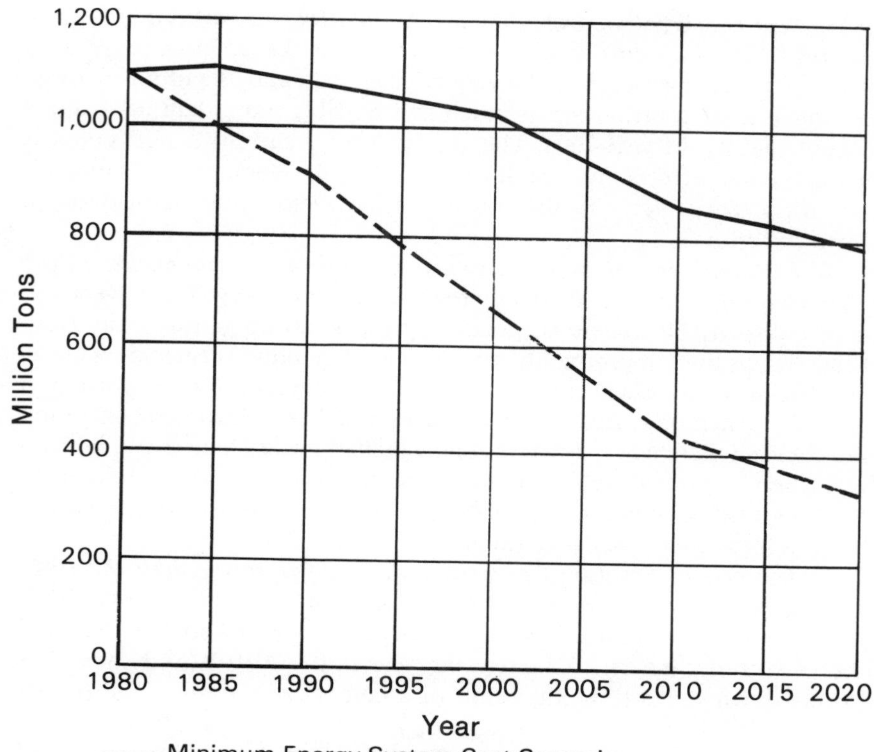

Figure 6-2. IEA Energy RD&D Strategy Project

per day (mbd), a large amount when compared with the fact that in 1980 all the IEA countries together imported approximately 22 mbd. Although it is not suggested that we are likely to actually reduce oil imports all the way to the lower level, it is clearly the direction in which IEA countries will want to move, and it demonstrates that new technologies can make very major contributions to reducing oil imports.

2. *The contribution of new technologies to oil-import reductions before the year 2000 may be inadequate unless their introduction is accelerated.* The possible savings of around 7 mbd by the year 2000 shown in figure 6-2 results largely from the availability of new technologies earlier and in greater amounts than in the business-as-usual case. In terms of the group policy objectives adopted by the IEA, technology acceleration thus becomes imperative, even if the actual savings is 4.5 or 6 mbd.

3. *New conservation technologies can be a very significant factor in meeting IEA oil-import-reduction objectives.* In the accelerated security scenario these technologies had a very large impact, and as a group, showed the potential of contributing as much and possibly more than new-supply technologies by the year 2000. The implication for current RD&D policy is that maximum efforts must be devoted to their development and commercialization, particularly in the transportation sector, and particularly in North America.

4. *The major growth in primary energy occurs in coal and nuclear power in virtually all countries.* These two forms of energy emerge from the analysis as indispensable. There is, practically speaking, no getting away from them. Coal is used to produce liquids, it is used in industry (including use as a feedstock for the chemical industry), and it continues to be important in electricity generation. But the largest category of new-electricity generation is in conventional nuclear reactors. In the limited-nuclear-power and limited-fossil-fuel scenarios it was found that limiting the growth of one resulted in the other making up part of the loss, and that severe problems would exist if the growth of either was constrained.

5. *The rate of growth of electricity slows down throughout the 1980-2020 period, but it continues to exceed that of total primary energy.* Electricity use is projected to grow at 3.1 percent per year between 1980 and 1990, but to decline to 1.7 percent per year in the 2010-2020 decade. The slowing down seems to be the result of a number of factors: GDP (Gross Domestic Product) growth slowdowns; a slowdown in switching to electricity from other forms of power; conservation measures taking effect; and new electric end-use technologies such as electric vehicles being only moderately successful in penetrating markets. The implications for RD&D are the need to consider whether all the electricity-producing technologies now under development need to be pursued, and the need to give greater emphasis to nonelectric technologies. For example, we may not need all the different types of nuclear reactors now being developed, and we may not need some of the renewable-electricity-producing technologies.

6. *Only some new renewable-energy technologies begin to contribute before the year 2000, and their maximum contribution potential remains very limited.* Residential and commercial solar heating and cooling, fuels from biomass, and geothermal (hydrothermal) energy all make contributions by the year 2000. However, even by the year 2010 renewable-energy technologies contribute only 3.4 percent in the minimum-system-cost case, 6 percent in the accelerated-security scenario, and 10 percent in the scenario in which renewable energy sources are pushed into the energy system regardless of cost. From this it is concluded that RD&D on renewable energy should focus on the more promising types already mentioned, insofar as the energy needs of IEA countries are concerned. Some of the renewable-energy

technologies, to be discussed, assume importance as hedges against the constraint of a mainline-supply technology such as nuclear technology. There are clearly non-IEA countries where renewable-energy technologies may have greater impact, and RD&D in IEA countries should not ignore these applications.

7. *Limitations on fossil-energy use would have major effects on economies and energy costs, and would increase oil and gas imports.* In the analysis, this scenario had the most serious results of any scenario analyzed. The indications were that there would be energy-supply shortfalls with resulting involuntary reductions in economic-growth rates, and sharp increases in oil and gas imports as well as in energy-system costs. Electric vehicles, fuels from biomass, wind power, and nuclear power were the main sources of energy that replaced coal in this scenario. RD&D should be directed at preventing such a future through aggressive environmental-technology programs, and through attention to the three main "hedge" technologies (solar, biomass, and geothermal power).

8. *Limitations on the use of nuclear power would require increases in all other forms of energy, and would lead to higher oil imports.* This simulation of the effects of continuing delays in nuclear-power programs showed that, rather than a few "hedge" technologies taking the place of a portion of nuclear power, a broad range of new, more costly technologies would be required. To minimize the chances of the realization of this scenario, nuclear-safety R&D as well as nuclear-fuel-cycle technologies must receive high priority.

Development of the Group Strategy for Energy RD&D

The results of the MARKAL systems analysis played an important but not an exclusive role in shaping what became the IEA Group Strategy. The first and most prominent element in the Strategy is the division of all generic alternative-energy technologies into four priority categories. This was done by applying the following criteria to each technology:

1. *The potential amount and timing of each technology's contribution to energy supply or to energy conservation.* If a technology was projected to make a high contribution (over 50 million tons of oil equivalent (mtoe) per year on average throughout the time period), then it was given a high priority ranking. If its contribution came early in the period, and if it produced liquid fuels, it was given extra credit. If the analysis showed that a technology would be a major contributor, but that because of long-lead development times, its major impact would only be realized toward the end of the time period, steps were taken to avoid penalizing such a technology (for example the breeder reactor) in the priority rankings.

2. *The degree to which today's state of knowledge makes it possible to estimate that the technology can successfully achieve the economic and technical goals assumed for it.* We can today be more confident that costs and technical characteristics assumed for a coal liquefaction plant built in the 1990s might, in practice, be realized, than we can be about cost assumptions for a wave-power-generating station built in the post-2000 period.

3. *The importance that certain technologies may have as hedge technologies.* Hedge technologies are defined as those which would be needed if one of the mainline energy technologies experienced a major failure or was severely constrained. The analysis of the limited-nuclear-power and limited-fossil-fuel scenarios provided insight as to which technologies would help to make up for the reduced availability of either nuclear or fossil-based power. Important hedge technologies were also given extra credit in preparing the priority rankings. Wind power was one of these.

Applying these criteria to the projected contributions of new technologies in the accelerated-security scenario, the scenario selected to form the basis of the Group Strategy, yielded the technology priorities shown in figure 6-3. The priorities are the most visible part of the Strategy. They represent a major step in international technology policy, for this is the first time that a group of nations have jointly developed, agreed upon, and promulgated specific priorities for energy technologies. They have thus achieved something that, as pointed out earlier in this chapter, individual countries have found exceedingly difficult, either because there is no policy interest in setting priorities or because the process is a political mine field.

The Group Strategy has various elements: *priority rankings* for energy technologies, *pursuit of an accelerated path* of RD&D and energy-technology commercialization, *indicative actions*, and attention to *environmental questions*.

That an accelerated path of technology development is a crucial part of the Strategy is to say that significant oil-import reductions are vital to the IEA countries, and that new technologies will only realize their large potential to reduce oil imports if there is a very major increase in efforts to develop and deploy them. In the real world this could mean a whole range of measures by which governments might encourage the use of new technologies such as coal liquefaction or heat pumps, before the technologies have become attractive investments because of the OPEC price for oil. The U.S. Synthetic Fuels Corporation is one such measure chosen by the U.S. government in 1980. Other administrations and other countries might adopt different measures.

The indicative actions (also shown in figure 6-3) are judgments by the IEA as to the most appropriate next stage of development for each type of technology. For instance, it is the IEA's view that breeder reactors are ready for demonstration, but not yet ready for commercialization. The value of

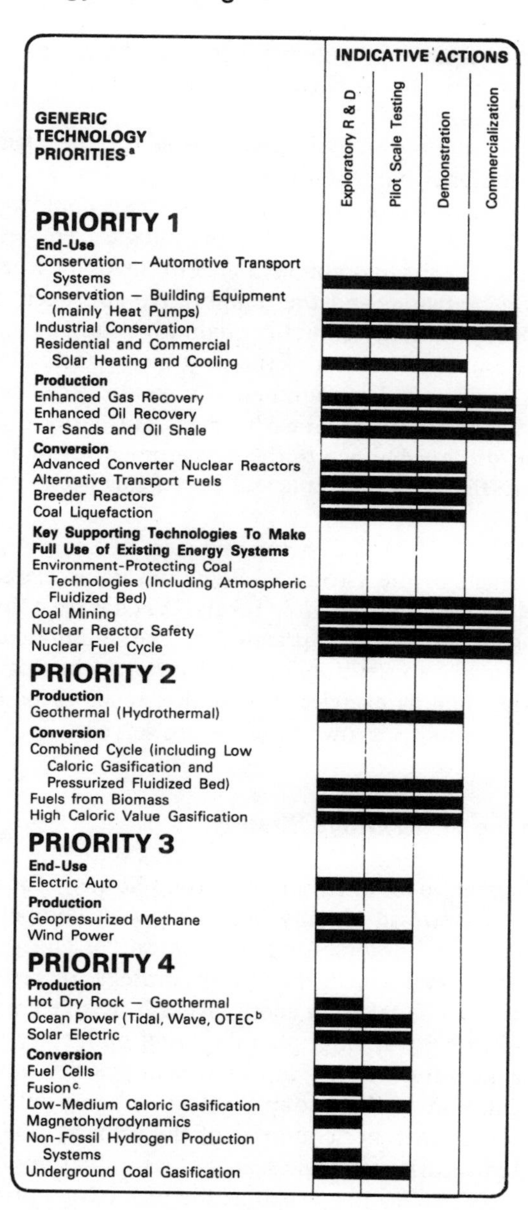

[a]The priorities apply to IEA countries as a group and not to any individual country.
[b]Ocean Thermal Energy Conversion.
[c]Fusion is ranked in Priority 4 due to its late availability (2020). However, taking into account its potentially large contribution and long lead time, sufficient efforts should be made to develop this technology.

Figure 6-3. IEA Energy RD&D Group Strategy

these indicative actions to policymakers may be increased when they are made more specific by application to the individual processes in each technological area.

During the systems-analysis preparation for the Strategy, one of the goals was to use environmental indicators in such a way that the effects of new energy technologies and new energy systems could be projected along with the market impacts of the new technologies. However, it was found that the state of the art had not advanced to the point where such a parallel analysis could be made, and the quantitative approach was dropped. Instead, the Group Strategy contains qualitative judgments on the different possible environmental effects of the new energy technologies. It also calls on countries to ensure that sound environmental-R&D programs are built into their R&D programs at an early stage. This latter point is particularly crucial, for it did not happen in the development of nuclear power, and it contributed to the difficulties nuclear power has been experiencing for the last ten years.

On May 21-22, 1980, the energy ministers of the IEA governments gave their endorsement to the Group Strategy and its accelerated path of technology Development. The next step for the IEA has been that of implementation. Putting strategies into practice at the multilateral level, whether economic assistance to developing countries or technological development, can be a rather hollow exercise. Realization tends to be a function of the degree to which countries' own interests are served.

Implementation of the Group Strategy

What is to happen to an international Group Strategy for energy-technology development in a world already brimming with competing notions as to which groups of technologies are best, safest, or cheapest? As far as the IEA countries are concerned, the Group Strategy, representing a consensus on the direction to be taken in energy RD&D, is an element to be added to governmental priority-setting, planning, and budgetary processes. It is intended to be an influential guide to national priority-setting.

Implementation of the Group Strategy is being defined by the IEA as the application by member countries of the primary policy assumptions in the accelerated-security strategic scenario, as well as application of the specific elements of the Strategy such as the technology priorities.

An important use of the ranking of technologies in priority categories is to indicate which technology should have first claim to funding. Priorities are not intended to be synonymous with size of programs. For example, developing a better heat pump will need far less money than demonstrating a solar-electric power plant. The priorities are simply intended to indicate which technologies should, in a resource-constrained environment, have their RD&D more fully supported than other technologies. One test of whether a technology is in fact being accorded first priority would be if one could say that all of the worthwhile RD&D needs necessary to advance that technology through all stages of RD&D with maximum speed are being fully supported. If progress is, in fact, project-constrained, that is, if there are no worthwhile projects left unfunded, then there is evidence that the technology is receiving first priority in a country's national RD&D planning.

The theory of applying technology priorities in a national context is relatively straightforward, but it is not immediately apparent how priorities based on the collective needs of a group of nations should be applied. How should national energy RD&D priorities be influenced by those of the Group Strategy? The Group Strategy priorities result from projections of energy impacts and they are therefore more relevant to the countries in which particular technologies have made significant impacts than they are to other countries (the principal exception being countries that would develop a technology for export).

For each type of technology the implications of the Group Strategy priorities will vary according to the group of countries for which that technology is important. For example, coal liquefaction appears in priority One because it is projected to make a high contribution to energy supply, but this technology will be used to a significant extent only by a limited number of countries. Australia, Canada, and the United States will use it because they have coal. Germany and Japan will use it because they need the liquefied coal. Thus, the priority designation will have different implications for different groups of countries, depending on the technology.

In short, what the priority One and Two technologies mean for individual countries is that the technologies for which a large potential has been demonstrated in the Group Strategy should be given top priority within national RD&D-planning processes in the countries showing large potentials. The Strategy also gives those countries that are leading in the development of a particular technology a notion of the scale of deployment that will be needed by a certain date if important oil-import reductions are to be achieved.

Thus under each technology area there will be a different grouping of countries, as few as four or as many as ten, who will be the main developers or customers for that technology, and who, therefore, are responsible for technological progress in those areas. These groups will need to develop their own specific strategies in order to know how much, if any, additional RD&D effort is needed to have the technologies ready in time.

In the case of priority Two, Three, and Four technologies, the message for individual countries is to examine their national priorities in the light of the Group Strategy conclusions concerning each technology, in order to make sure that their national priorities are in reality justified by future national energy needs.

Certain technologies may be of major importance to the energy future of one country, and therefore deserve top priority within that country's RD&D program, whereas on an IEA-wide scale their projected impact does not justify top priority. The Group Strategy recognizes that some technologies may be of unique national importance, and in such cases the Group priorities would not be relevant. One example of such a technology is the process for making gasoline from natural gas. This technology is particularly attractive to New Zealand, which has large amounts of natural gas, but has to import refined gasoline.

Acceleration

An accelerated path has been recommended in the Group Strategy and specifically approved by the energy ministers of the IEA. Acceleration is measured against a business-as-usual approach, or the pace of technology development prevailing up to 1978 (when the analytical assumptions were frozen). In the Group Strategy, acceleration means significantly increased RD&D and commercialization efforts, and in the systems analysis its effects were simulated by assuming that technologies would be available earlier and in greater amounts than in the business-as-usual case. It is not a difficult matter to assess whether or not acceleration is taking place, both in general and for individual technologies.

There are, nevertheless, some elusive questions to answer with respect to acceleration. How much acceleration should be sought over the next five to ten years? How should one judge whether some technologies (such as certain nuclear technologies) have already been sufficiently accelerated and should not receive increases similar to those accorded to technologies that have been relatively neglected over the past ten years? What are the R&D infrastructural and financial limits to acceleration? RD&D policymakers in each group of countries interested in a particular technology will have to address such questions, and they will receive encouragement to do so from their respective treasuries, finance ministries, and offices of management and budget.

Monitoring Implementation

It might be asked how the IEA keeps track of the way countries are using its Strategy, how it measures whether we are moving fast enough in getting the high-priority technologies ready for the marketplace. In the case of the

IEA, there already exists a process for monitoring the progress made by member countries in implementing joint policies. Each country is subjected to a review or examination, conducted not by international civil servants, but by officials of other countries selected for each review. They scrutinize the recent actions of the country under review, and prepare a short critical report on their findings. The objective is to put pressure on the government concerned to move in the agreed direction. These reports are made public with the expectation that their conclusions will receive exposure in the media, and that pressure on policymakers will result. The IEA plans to use its ongoing review process as one vehicle for monitoring implementation of the Group Strategy.

Environmental Considerations

The Strategy emphasizes the importance of incorporating RD&D on environmental work into the early stages of the R&D process. The IEA countries, collectively and individually, will need to assess whether any specific new programs are needed to prevent delays in commercializing the major alternative technologies because of adverse environmental impacts. At the same time they will be able to monitor the actions taken by individual countries through their ongoing national-program review process.

Implementation in International Cooperative Technology Programs

Another area where the influence of the Group Strategy may be felt is that of international (bilateral and multilateral) projects to develop new energy technologies. The IEA has, over the last four years, built up a program of international energy-RD&D projects in nearly every area of energy technology except nuclear fission, which is covered by the OECD's Nuclear Energy Agency. As of early 1981 the IEA program included close to fifty projects with a total value of over $650 million. The Strategy will be used to shape priorities in the arena of multilateral technology development.

Total Public and Private-Sector Energy-RD&D Efforts

Although the Group Strategy may have greater potential to influence governments by virtue of the fact that the IEA is made up of governments, the Strategy is derived from projections of *national* energy-system requirements, and as such is based on the aggregate of government and industry actions to develop and commercialize new technologies. Accurate estimates of the energy-RD&D expenditures of industry are difficult to obtain. In its country-review process the IEA attempts to collect such data, but has had very partial success to date. Information at hand indicates a total industry effort in 1979 of approximately $4 billion as compared with government outlays that year of $7 billion. As one would expect, industrial expenditures are directed at technological developments expected to have short-term payoff. However, governments do influence the direction of industry RD&D spending, through grants, tax incentives, and cost-shared projects. To the extent that governments are generally following a path similar to that of the Strategy, and to the extent that industry's own perception of the long-term technological needs of the industrial countries resembles that in the Strategy, it may have some impact on private-sector R&D directions.

The Summit-Level Initiative and Alternative Technologies

What has thus far been described is an international initiative to guide and stimulate the development of alternative-energy technologies through the established IEA mechanisms of cooperation on technology policy. In 1979 action was taken at the highest levels of government to advocate those new technologies likely to make large energy contributions before the end of the century. These initiatives have not received much publicity, lacking the media appeal of actions taken to cope with the most immediate economic problems of the day. Nevertheless, the careful student of summit communiqués will find evidence of a U.S.-inspired initiative to push for alternative-energy technologies in the final pronouncements from the Tokyo and Venice economic summits in 1979 and 1980.

The International Energy Technology Group (IETG)

In June 1979 the Tokyo Economic Summit (attended by Canada, France, Germany, Italy, Japan, the United Kingdom, and the United States), decided that answers were needed to the questions: are the present efforts of the public and private sectors to commercialize alternative-energy technologies adequate, and, if not, what more needs to be done at the top international and national political levels? This marked the first time that alternative-energy technology development had gained the attention of heads of state. The Tokyo Summit set up a special working group of top government officials (the IETG), chaired at the undersecretary level, for the purpose of quickly assessing the present situation and reporting to the following year's summit on new technologies that were basically ready to be commercialized and required government action to speed their entry into commercial use. Technologies that industry itself would commercialize were to be excluded from the list of candidates for special government encouragement.

In its report to the Venice Economic Summit of June 1980, the IETG said that the financial institutions and mechanisms existing in the Western industrial countries were adequate for the task of commercializing alternative technologies and that no new institutions were needed. But it singled out six areas of new technology where the private sector was unlikely to move quickly into commercial deployment either because technologies had not been technically proven on a commercial scale or because the economics of commercial-scale operation were too uncertain in light of the very large financial requirements for even a single plant ($1-4 billion). The six areas were:

Coal liquefaction;

Tar sands, oil shale, and heavy-oil upgrading;

Fuels from biomass;

Coal gasification;

Liquid fuels from natural gas; and

New coal-combustion technologies.

The IETG recommended to the Venice Summit that it set in motion a two-phase plan for commercializing technologies in these areas. The first phase would cover the decade up to 1990, and would pull together all of the projects planned in the participating countries involving commercial-scale plants. The idea of the first phase was to obtain an accurate picture of the production capacity that was likely to be in place by the end of the 1980s. This picture was to be built up on a project-by-project basis. The second phase would cover the decade 1990 to 2000, and would take as its departure point the inventory of alternative technologies in the six areas producing energy as of 1990. It would then project the likely market penetration of each of the six technology areas through the 1990s. With these projections in hand, governments would have a realistic picture of what they could expect new technologies to contribute based on actual plans, and they would have a basis for deciding whether any of the technologies needed special action to speed up their commercialization.

The High Level Group for Energy Technology Commercialization (HLG)

Having made its report[4] to the 1980 Venice Economic Summit, the IETG went out of existence. However, its recommendations were all approved at the Venice Summit, which then set up a new body, the High Level Group for Energy Technology Commercialization (HLG), to carry out the actions that it had approved. The report submitted by the HLG to the economic summit in Ottawa in the summer of 1981 contains a unique set of data from which it will be possible to obtain an authoritative picture of the energy supply that might be obtained from new energy technologies by the 1990s and the year 2000 based on an exhaustive survey of the plans of both the public and private sectors. More important, the report analyzes the constraints that are delaying the commercialization of alternative-energy technologies, and thus provides a basis for governments to clear away those obstacles that are within their power to influence.[5]

It remains to be seen whether the elevation of energy-technology policy to the world of summitry will have the intended effect of cutting through bureaucratic delay and industrial caution. The objective is certainly a worthy one. The current problem of those responsible for long-range technological development is to encourage the institutions who hold the purse strings—corporations and banks—to make investment decisions on large energy projects when the payoff is uncertain not only because of technical and economic risks but also because of changing governmental regulations. The summit is one level at which measures could be encouraged that would stimulate investment by industry.

The initiatives that have been summarized here are taking place at the intergovernmental level. To these must be added a very active set of international relationships among private companies that have flourished without and sometimes despite government involvement.

At the national level, particularly since late 1979, some very major programs have been planned and launched in Germany, Japan, and the United States with the objective of bringing about the creation of entire new industries devoted to alternative-energy technologies. These include the development of new energy-conservation and supply technologies, though the latter receive priority in the international initiatives mainly because of their very high unit costs. It was the task of the HLG to provide perspective for these national programs so that the people responsible for their direction in the public and private sectors may gain a broader view of whether the results awaited and the speed with which they are being achieved are commensurate with the very grave and taxing energy future faced by the Western industrial countries.

Conclusions

This chapter has approached the subject of alternative-energy technologies from the view of international efforts at the government level to accelerate those alternative-energy technologies that might contribute most to reducing our dependence on imported oil. Central to the entire process is the role of the private sector, where the decisions to manufacture and deploy these technologies will be made. The extent to which governments should involve themselves in that part of the technology development process concerned with moving technologies out of their RD&D stages and into commercial use is a matter of considerable debate. This, however, is an issue that must first be resolved within national borders. We are at the very beginning of a learning curve, for the postwar experience of governments' role in the commercialization of nuclear power does not hold many useful lessons for the new alternative-energy technologies, except perhaps in learning from past mistakes.

International efforts do make sense because the energy problem faced by the oil-importing countries is a shared one. But these same efforts will have to take into account not only different technology needs (North America and Australia have needs for liquid fuels of a very different nature from those of Europe), but also different political philosophies with respect to government's role in the commercialization process. Nevertheless, it is clear from what we know today that the technologies with very large unit size (one commercial-size coal liquefaction plant is estimated to cost $4 billion), and that are still unproven, are very unlikely to be built by private industry, because industry will not invest very large amounts of capital in technologies that are unproven at commercial scale when it can obtain a satisfactory rate of return from less risky investments.

At the international level, the most useful developments will probably be those that encourage the early construction and operation of enough power plants to prove the technical and economic viability of those technologies that could realistically be important in the 1990s. Such plants take five to seven years to build. Whether we will need large numbers of them in the 1990s cannot be accurately forecast today. But if we have not proven the technologies by the end of this decade, the option of alternative sources of liquid and gaseous fuels may not exist for us in the 1990s. Once technologies have reached a technical and economic "take-off" stage, it will be up to governments to decide whether to rely on market forces or to pursue more interventionist policies to accelerate use of new technologies for reasons of energy security. In either case it will be essential that governments encourage stability in regulatory processes, equitable treatment for foreign investment, and access to prevailing world energy prices.

The fact that this chapter has focused entirely on the technology needs of the industrial countries should not obscure the very urgent, but quite different needs of the oil-importing developing countries. These are being addressed by a number of international and regional organizations with varying success, and as much space again would be needed to describe those activities.

Notes

1. These and other data on government energy R&D budgets used in this chapter are taken from the annual publication of the International Energy Agency, *Energy Research, Development and Demonstration in the IEA Countries, Review of National Programmes* (Paris: OECD, 1980).

2. Participating governments are Australia, Austria, Belgium, Canada, Denmark, Germany, Greece, Ireland, Italy, Japan, Luxembourg, Netherlands, New Zealand, Norway, Portugal, Spain, Sweden, Switzerland, Turkey, the United Kingdom, and the United States.

3. For a more complete description the reader is referred to *A Group Strategy for Energy RD&D* (Paris: IEA, 1980).

4. The report of the International Energy Technology Group (Paris: OECD/IEA, 1981).

5. The report of the High Level Group is summarized in Niels de Terra and Willem Smit, "Synthetic Fuels: OECD/IEA High Level Group Proposes Plan for Commercialization," *OECD Observer*, no. 111 (July 1981), pp. 27-29.

Part II
Western Energy Policies after the Second Crisis

7 British Energy Policy

N.J.D. Lucas

This chapter begins with a sketch of the institutional structure of the energy sector in the United Kingdom. It then reviews the development of energy policy in the United Kingdom since World War II and describes energy policy as it is today. The role of the United Kingdom within the European Communities and the changes brought about by the second oil crisis following the Iranian revolution are also discussed.

Structure of the British Energy Sector

The principal institutions involved in the energy sector are the political institutions, the government departments, the nationalized industries, the private oil companies, and the consumers. The most important government departments are the departments of Energy and the Treasury, but almost all government departments have some interest in energy; the Foreign and Commonwealth offices, and the offices of Industry, Transport, and of the Environment have perhaps the most substantial secondary involvement.

The principal state energy agencies are the gas, electricity, and coal industries (nationalized after World War II); the United Kingdom Atomic Energy Authority (UKAEA), established in 1954; and the British National Oil Corporation (BNOC), established in 1975 to represent state interests in the North Sea. The British Gas Corporation (BGC) has a monopsony on gas for fuel (not petrochemicals) and a monopoly on fuel transmission and distribution; the form of the industry has evolved considerably since nationalization, but with a constant tendency toward centralization of control. The Electricity Supply Industry (ESI) is composed of the Central Electricity Generating Board (CEGB), with a monopoly on high-voltage-electricity distribution and twelve area boards responsible for retail sale of electricity to customers. The Electricity Council is the forum where the general policy of the electricity-supply industry in England and Wales is formulated. The generation and distribution boards are fully represented on the Council, which also has independent members, one of whom is the chairman. The ESI has evolved considerably since nationalization, with a not-quite-continuous trend toward centralization. Several attempts have been made to reconstitute the industry into a single corporate body; so far they have not been sanctioned politically. In the immediate past there has been a de facto centralization arising from the increasing tendency for the

Electricity Council to exert its authority. The National Coal Board (NCB) has a monopoly right to the mining of coal; again there has been a continuous centralizing tendency since nationalization, more, in the case of the NCB than the BGC and the ESI, through internal management reforms than by statute. The BNOC has complex and rapidly changing privileges in the North Sea that are discussed later in the chapter.

The Political Structure

It is important to have some idea of how these technical agencies fit into the political structure.[1] The United Kingdom is a constitutional monarchy and in principle is ruled by the Crown; convention has established that in practice the Crown does what ministers say, so it is ministers collectively that initiate and enforce policy. But the source of their powers is the sanction and legitimacy conferred by Parliament. For the purposes of energy policy the relevant powers are statutory powers. Statutes confer powers and duties on ministers or on existing or new agencies. In return for its sanction Parliament demands the right to monitor the exercise of the powers confirmed. Therefore there is always specified in all statutes a minister responsible to Parliament for the acts of central government agencies. Parliament has also created Select Committees of its members which scrutinize the activities of government departments and inform the Assembly.

Public corporations are created in areas where it is desirable for an activity to be under public ownership, and where the exigencies of its day-to-day business require a boldness and entrepreneurial spirit that are not felt to be characteristic of government departments. The conflict between the desire for autonomy within the energy enterprises and the need for the government to affect the activities of the enterprises is the principal force governing the relationship of government and state enterprise.

Conferral of powers on agencies outside government departments has a long history in the United Kingdom. As life has become more complex and technical and as the state has become more involved in industrialization and in correction of externalities, so the delegation of powers has required that the effects of the exercise of these powers be rigorously monitored, and the conflict between autonomy and control has been exacerbated.

The formal development of the relations between government and nationalized industry is described later, but broadly, the original idea was that ministers should determine general policy, and that the enterprises should then conduct their affairs, within those constraints, as commercial bodies. This idea is rooted in the notion that the activities concerned are non-political (such as gas and electricity production or coal mining) and must be governed by commercial logic. This has never been true; therefore ministers have been tempted to intervene clandestinely, and civil servants have surreptitiously persuaded state enterprises to modify their intentions. This process has confused responsibility and power. One of the great successes of the Select Committees has been the revelation of this process by the Select Committee on Nationalized Industries[2] and more recently by the National Economic Development Office.[3]

Up to this point we have assumed that government departments act, as they are supposed to, as agents of their ministers. But in practice the civil service is the most uniformly able body in the United Kingdom and it has its own ideas. Ministers change very quickly (about one year to eighteen months is the typical duration of a minister's responsibility for energy), and they have little or no technical support outside of their department. In the case of the various ministers of Fuels and Power or of Energy, they have tended to be men either on their way up or on their way down, but in any case with other things on their minds. Senior civil servants also change often (though not as often as ministers) but the collective view of the department is unaffected. Much of the arm-twisting applied by civil servants to public enterprise is directed to ends that each department wants; this is known by the technical agencies and increases their resentment.

Evidence for the dominant role of the civil service in policymaking comes from ministerial anecdotes,[4] and also from the astonishing doctrinal continuity between the policies of different political parties over matters of substantial political significance, such as the rundown of the coal-mining industry in the early 1960s, or, more widely, the attitude toward the control of nationalized industries.

A very important force in the formulation of departmental policy is the Treasury; indeed it is hardly possible to overstate its influence. All aspects of departmental policy are constantly checked with the Treasury, giving it an unremarked but extremely efficient centralization of control. The process is reinforced by the official committees that shadow ministerial meetings, prepare briefs, and operate policies: these committees are constantly influenced and informed by the Treasury. The respect for the Treasury stems partly from its traditional role as final coordinator, partly from the fact that promotion depends strongly on Treasury favors (so that loyalty to the Treasury is more likely to pay off than loyalty to a fleeting minister), and partly from the fact that top civil servants have usually worked at the Treasury and have been imbued with the requisite sense of loyalty. The dominance of the Treasury is all the more powerful in that it is embedded in the civil service machinery one step from direct political control.

The increasing recognition of the political nature of technical decisions that has bedeviled relationships between government and state industry has been paralleled by greater activity from a third, and increasingly eloquent, element—the consumer. Consumer councils for the coal, gas, and electricity industries exist to provide an organized representation of the interests of those consumers, mostly domestic, who have to buy the final product and are otherwise unrepresented. In the past their main activity has been to protest price increases. Independent of the consumer councils there are several causes with a strong interest in energy policy; increasingly the consumer councils, the interest groups, research groups at universities, and to some extent local authorities are making formal or informal alliances to produce mobilization of interests affected by new energy developments. This must inevitably further strain the concept of state enterprises as bodies that should ideally operate by commercial logic, and must make more complex their relationships with government.

The Evolution of Relations of Government and Nationalized Industries

When the coal, gas, and electricity industries were nationalized, the fundamental financial instruction given them was to pay their way without being a burden on the taxpayer or exploiting any monopoly position. This was typically expressed in the statutes as a requirement to break even, after payment of interest and after proper allocation for depreciation; the statutes also required the industries not to discriminate unduly between different classes of consumers. These requirements implied minimum returns from the industries and carried a risk that resources that could be used more effectively elsewhere would be taken up by state enterprises. Government White Papers in 1961 and 1967 sought to provide the industries with positive financial objectives.[5] The thrust of this development was to use a Test Discount Rate (TDR) for project appraisal (first set at 8 percent and later raised to 10 percent), to align prices on long-run marginal costs (LRMC) and to achieve financial targets covering several years. The main form of target has been a percentage return, before interest, on the average net assets employed by the industry.

The philosophical problem with this formulation is that it overdetermines the model of the industries' financial performance. If prices are aligned on LRMCs and a fixed TDR is used for investment appraisal, then the financial target follows. The academic escape from this is to say that the financial target dictates the overall price level of the industry and that LRMCs determine relative prices such as the capital and unit components of electricity prices, the geographical relativities in coal production, or large and small sales of gas.[6] In fact, more substantial practical problems developed. In the first place, LRMCs are too difficult to determine and subject to too much uncertainty to be a useful practical guide. For example, in the coal industry a long debate developed about which coal is marginal; if the maintenance of present pits is considered indispensable on social grounds then the incremental cost is the cost of brand-new capacity with high productivity; but if the old capacity is closed the decremental cost is the cost of the most expensive coal. Add to this the concept of resource costs, and the marginal cost of coal becomes anything from the most expensive to the least expensive coal, modified by an arbitrary factor representing the resource cost of labor. In other cases prices are largely set by trading conditions—coal in the 1960s, oil prices for the BNOC today. Alignment on LRMCs is undesirable in these circumstances.

For these and related reasons, in different industries the financial targets took precedence. The principle of financial targets was then undermined by price restraint in the first half of the 1970s, which forced the gas and electricity industries into deficit and caused them for the first time to accept subsidies to cover the loss. The relationship between TDR and LRMCs also turned out to be less clear than originally expected because much investment was of a form not susceptible to this sort of appraisal (it was essential for safety or security). As a guide for selection among competing projects the TDR evidently must work, but it has not been found to provide nor stimulate the development of an adequate way of relating the cost of capital to the industries' financial objectives.

The government, aware of the inadequacies of the existing system and in possession of the severe criticisms of the National Economic Development Office (NEDO), which asserted, "Our enquiry has left us in no doubt that the existing framework of relationships, developed under governments of both main political parties is unsatisfactory and in need of radical change,"[7] issued a new White Paper replying to the NEDO report and containing new proposals.[8] This White Paper is essentially reformist and concentrates on correction of the inadequacies of the financial obligations of the industries. The use of a TDR for project appraisal is replaced by an obligation to achieve a required rate of return (RRR) on new investment as a whole. It was left to the industries to design procedures for appraisal that would achieve the overall return. This gives the industries considerably more discretion in selecting projects according to their own political priorities. The present judgment of the government is that a RRR of 5 percent in real terms before tax is appropriate.

The industries also have, or will have, a financial target expressed as a return before interest on the average net assets employed by the industry; the level of the target will depend on the sectoral and social objectives set for the industry and on the earning power of its existing assets.

The recent preoccupation of Her Majesty's government with public spending has been partly expressed as a constraint on the nationalized industries' external-financing requirements. A limit is set for each financial year on the permitted resort to external finance. This seems again to overdetermine the model, because once investment criteria and financial targets are set, the requirement for external finance follows. The effects of cash limits, if they continue, will be to restrain investments or to increase prices. There is some doubt, for example, whether the Electricity Council will be able to finance the planned nuclear expansion within its present cash limits. The overdetermined constraints this time are more serious than in the 1967 White Paper because they are less easily circumvented. Conservative ministers who in the past have been instinctively hostile to state ownership are beginning to realize that restricting investment in natural-gas pipelines, coal mines, or nuclear power is quite different from cutting public expenditure on services. A possible solution is that new methods of finance will be required that will permit private capital to be attracted to specific projects. The important consequences for energy policy of these means of financial control are: (a) prices tend to be determined by historic costs through a financial target relating to historic assets; and (b) investment policy is determined not by profitability, but by arbitrary restrictions on borrowing.

Historical Evolution of Energy Policy

The time since World War II can be divided broadly into four periods separated by major changes in the dynamics of the British energy sector. From World War II to the late 1950s energy supply was based on coal and supplies were scarce. The gas and electricity utilities were dependent on the coal industry; the relationship among the newly nationalized industries was one of interdependence. In 1957 competition from oil began with a severity not really foreseen by anyone; gas and electricity abandoned the coal industry; direct and indirect competition with oil in all markets became severe. In the mid-1960s the highly competitive nature of the fuel sector was enhanced by the arrival of indigenous natural gas, increasing the pressure on coal and nuclear power. It was a period when policy objectives were clear, but it did not last long. Much of the direction of policy was reversed by the oil crisis of 1973, which exposed a contradiction between the existing policy and real long-term needs. The conflict between short- and long-term needs was further sharpened by the discovery of North Sea oil and natural gas.

From the End of World War II to 1957

Energy supply at this time was based almost exclusively on coal; in 1950 primary-fuel consumption in Britain was some 140 million tons of oil equivalent (mtoe), of which 85 percent was coal (200 million tons of coal). By 1957 coal consumption was at about the same level, but represented only

75 percent of total primary-fuel consumption; in the intervening period fuel shortages had come to an abrupt end and incremental demand had been met by oil. Even so, the other nationalized industries still depended heavily on coal; in 1957 coal made up 94 percent of the primary-fuel input to the electricity industry and coal or coal products made up 93 percent of the feedstock for the gas made or bought by the public gas industry. The relationships among the coal, gas, and electricity industries were determined by bilateral discussions and little coordinated planning. Even by 1958 the gas industry was indicating that it would take 27 millions tons (mt) of coal in 1965 and 30 mt in 1970; in fact it took 18 mt in 1965 and 3 mt in 1970. Earlier, in 1950, the government had urged the coal industry to increase output to 240 mt and the Federation of British Industry had argued that the objective was too small and had proposed 270 mt. The coal industry invested to meet the expected demand and was left with the burden of the cost of its unused capital and human investment. The coal industry was legally obliged to try to supply coal of the appropriate quality, but there was no obligation on its customers to take it. The sunk costs were paid entirely by the NCB, spread over a smaller output than expected, leading to higher prices and a vicious spiral of decline. A retraction of the industry was of course necessary, but there can be no doubt that uncoordinated planning exacerbated the consequences of the penetration of oil into the energy market.

1957 to 1965

The principal element of energy policy in this period was the defense of coal and the protection of nuclear power against oil. The defense of coal was undertaken exclusively for social reasons; the NCB argued that oil was insecure, but that consideration never swayed the government; nuclear power was protected as an infant industry.

The effect on the coal industry was catastrophic; production fell from 224 mt in 1957 to 193 mt in 1961; undistributed stocks rose to 37.5 mt. The six main areas of sale were simultaneously under pressure: electricity, railways, steel, gas, industrial, and domestic. Steel was a declining market because of the low growth in output and rapid increase in efficiency; railways were rapidly changing to diesel; town gas was made more cheaply from reformation of liquid natural gas (LNG) or naptha; domestic use was hit by slum-clearance schemes, replacing individual dwellings with high-rise buildings with gas or oil heat, and later by the Clean Air Act; industry was changing to heavy fuel oil. Only electricity generation was a growth market and even that was subject to competition from heavy fuel oil and nuclear power, but as another nationalized industry the electricity industry was susceptible to political pressure, and the NCB went to great lengths to maintain or increase coal use through political influence. The relationship among the enterprises underwent a marked change; competition replaced interdependence.

Contemporary with the penetration of oil was the development of nuclear power. The first nuclear-power program announced in 1955 and 1957 comprised nine stations of the Magnox type (4.5 Gigawatts). It is probable that these stations have all been strictly uneconomic although it is a difficult matter to assess because inflation has eroded the burden of the capital costs of these stations, costs which were at the time startling. The means by which the effects of inflation are quantified can influence the judgment. The principal reason for such a large program was the diversity of engineering consultants who all wanted to be involved, and the size of the program was determined by the U.K. Atomic Energy Authority (UKAEA) and the government. The CEGB, which was to have the responsibility of operating the stations, had not been consulted about the size or choice of technology.

During this period the gas industry greatly reduced its dependence on coal and began to diversify its supplies. By 1965 only some 45 percent of the gas distributed by the industry was based on coal. The rest was mainly made from the reformation of naptha, but significant quantities were also made from LNG imported from Algeria. Although methane is a less suitable feedstock for reformation than is naptha, the alternative supply gave the gas industry some negotiating advantage with the oil companies.

By the end of this period, therefore, the previous cozy coexistence of the nationalized fuel enterprises had been shattered by the penetration of oil. The costs of change had accrued almost entirely to the Coal Board (NCB).

1965 to 1973

Gas in the British sector of the North Sea was first found by the British Petroleum Company (BP) 45 miles east of the Humber in the autumn of 1965. In the same year the second nuclear-power program began with the receipt of bids for the Dungeness-B reactor. In 1967 the government published a White Paper on fuel policy to assess the balance between the four available primary fuels and to set the framework for the most beneficial development of indigenous supplies. Another period of rapid change began.

The first licenses for drilling in the North Sea were granted in September 1964. The initial response came mainly from the international oil companies; the British Gas Council participated with Amoco, and operation was left to the American firm. Labor-Party pressure to increase public participation in exploitation led to a preference for British interests in the 1969 round of licensing in which the Gas Council participated as an operator in its own right; the same pressure assisted the NCB in persuading the government to modify their statutes to permit them to drill for gas, thereby drawing on their considerable geological expertise; the NCB also participated with U.S. oil companies.

The major questions of policy were how to appropriate the considerable rent generated and how rapidly to deplete the fields, considerations which also affect the price of gas. Four methods of appropriating the rent were considered:[9]

1. auctioning of licenses—this was thought unpredictable and subject to collusion;
2. setting up a special state enterprise as the sole licensee and capturing the rent through a financial target set so as to pass the rent to the Exchequer—this was thought impossible in the time available;
3. a tax or duty on North Sea gas—it was thought the oil companies would find a way around this; and
4. attribution of monopoly purchase power to the Gas Council so that it could buy near cost and pass the rent on to the consumer or the state.

The last option was chosen. (It is interesting to note in passing that when it came to design, North Sea oil policy was a mixture of all four options, employed in different proportions.) Exploration for and extraction of natural gas in waters under British control is governed by the Continental Shelf Act of 1964; the minister of Fuels and Power (later of Energy) has powers under this act to deny the exploring companies the right to sell gas for fuel to any purchaser other than the Gas Council, providing the council offers a reasonable price. It therefore falls to the minister to determine a reasonable price.

In the matter of depletion rates the oil companies, with their near-monopoly at the time on expertise in reservoir engineering, had the better of the negotiations. They quite correctly persuaded the Gas Council and the government that there is an optimum depletion rate for a given field; this rate gives the lowest-cost gas and there is a substantial cost penalty in moving away from that rate. The most satisfactory method of controlling depletion of gas is by timing of exploration and production of successive fields. The fields in this case had been found and it was not thought reasonable to delay production in specific fields. The inevitable conclusion was that fields should be mined at their cumulative optimum rate of about 3,000 mcfd (million cubic feet per average day). This was about three times the volume of the town gas industry. The authors of the 1967 White Paper concluded therefore that "the greatest gain to the economy will be obtained by a policy of rapid absorption with most of the gas going to premium markets (where it would largely be displacing oil) and some bulk sales. . . . This will mean some displacement of coal by natural gas in the bulk energy market, but the coal markets that will be lost in this way would in all probability fall to oil if natural gas were not available.''[10]

The second nuclear program, announced in 1965, was for 8 GWe; only 6 GWe were actually ordered. The program was based on the advanced gas-cooled, graphite-moderated reactor (AGR). In principle the orders went out for commercial bids, but the UKAEA, alarmed at the prospect of an American Pressurized Light-Water Reactor (PWR) winning, connived with one of the consultants to propose an apparently lower-cost design; this design was accepted for Dungeness-B and the subsequent orders (for Hinkley Point, Heysham, and Hartlepool) went to AGRs. Dungeness-B has been fifteen years in construction, there have been serious construction delays with the other sites, and to date they have had a poor operating record. The unkindest cut, from the point of view of the NCB, was the nuclear station at Hartlepool, built on the Durham coalfield.

The effect of the entry of indigenous natural gas into the energy market was to exacerbate the already high level of competition; the second nuclear program has so far had no such effect, because the delays in implementing it were so long. The authors of the 1967 White Paper expected gas largely to displace oil in premium markets and to displace some coal in bulk markets. In fact, it mostly displaced coal in the domestic market (a premium market) and seems to have restricted electricity growth. The unforeseen effects on electricity sales caused acute overcapacity; the White Paper foresaw an installed generating capacity of 73 GWe in 1972 for a maximum demand of 61 GWe, compared with an installed capacity of 43 GWe in 1966 for a maximum demand of 40 GWe. Maximum demand in 1970 was 50 GWe. Although not all the planned generating capacity was actually built, and much of the system was built late, the system still possessed a 40 percent excess capacity with all the associated financial consequences. This is the second example of the underutilization of existing capacity following the introduction of a new energy source.

The authors of the 1967 White Paper foresaw coal production falling to 120 mt in 1975 (in which they were roughly correct) and 80 mt in 1980; this later figure was removed from the final version of the White Paper, but given out in Parliament.

The main error that underlies this White Paper is of course the belief in continued cheap oil. The chairman of the National Coal Board at the time recounts that a senior civil servant said of OPEC ''You can forget about them. They will never amout to more than a row of beans.''[11] It is certainly true that the same sentiment, differently expressed, appears several times in the White Paper.

The pattern described here of rapid penetration of natural gas into the energy sector at the expense of coal and electricity, continued to be considered acceptable until the oil crisis of 1973. The changes brought about by that event and the contemporary development of oil in the North Sea are discussed next.

1973 to the Present

Internally, the principal effect of the OPEC price increases in 1973-1974 was to upset the competitive balance among energy forms, again greatly in favor of natural gas. The crude-oil price increase was transmitted directly to the product market; the miners did not hesitate to extract the freshly created income almost completely in the form of wages, causing the price of coal to follow closely that of heavy fuel oil; electricity prices, being dependent on the prices of oil and coal, rose rapidly; and only the gas industry, protected by its monopoly on North Sea resources and favorable supply contracts, was unaffected by costs. From the OPEC crisis to 1977, electricity prices increased by 150 percent, gas prices by 75 percent. The change in competitiveness caused the domestic-market share of electricity to even out at about 20 percent and then to gently decline; the share of gas in the period rose from about 30 percent to nearly 45 percent. This growth took place mainly at the expense of solid fuels, whose share of the domestic energy market declined from about 35 percent to 25 percent.

In the two years following the energy crisis of 1973 the British government made several decisions:

1. it endorsed a program of investment to regenerate and expand the coal industry;
2. it took powers to control the development of offshore oil resources;
3. it proposed a program of 4 GW of nuclear power based on the Steam Generating Heavy Water Reactor (SGHWR); and
4. it announced a preliminary program of conservation of energy.

This strategy was a response to the underlying problem that coal, energy conservation, and nuclear power were now seen as important long-term options that had to be preserved or constructed during the medium term when indigenous supplies of oil and natural gas were plentiful. This conflict most underlies the development of energy policy in this period.

One of the first actions of the Labor government on taking office in March 1974 was to set up a tripartite group consisting of government, unions, and the NCB to examine coal-industry problems. The group endorsed the general aim of the NCB's Plan for Coal to provide 135 mt by 1985, 42 mt of which would come from new investment. New investment in fields such as Selby and Belvoir was expected to have much higher productivity than average, about 10 metric tons per man-shift (t/ms) as opposed to an average of 2.2 t/ms in 1971 (and 1.25 t/ms in 1957). The reserve base at existing pits was estimated at 4,000 million tons (mt); the discovery rate at potential new mines was 500 mt per year; and about 45,000 mt were estimated to be economically recoverable, that is, more than 300-years supply at current rates. The cost of the Plan for Coal was estimated at £3,000 million at March 1976 prices. The long-term objective for the industry was 170 mt by the year 2000.

The first substantial discoveries of oil in the British sector of the North Sea were made in the early 1970s. The Labor government considered it indispensable to exert control over this national asset and to ensure that a fair proportion of the income went to the nation. In 1974, following the creation of an enormous income by the OPEC price increase, the government undertook a fundamental review of policy. Arrangements were made to control offshore development and production and to share the income by participation and taxation. The legal instruments for this are contained in the Petroleum and Submarine Pipelines Act of 1975 and in the Oil Taxation Act of 1975.

The Petroleum and Submarine Pipelines Act, and some earlier legislation, enables the minister of Energy to regulate the rate at which fresh territory is licensed for exploration; to control development and production programs; to delay the use of production capacity and to require within limits that production from existing capacity should be reduced. These are comprehensive powers. An oil company, or group of companies, must submit development plans to the Department of Energy; before development is approved by the minister, the department analyzes alternative schemes. The legislation therefore permits the minister to control short-term or long-term discrepancies of any sort between company and national interest.

The other main purpose of the Petroleum and Submarine Pipelines Act was to establish the British National Oil Company (BNOC) as the principal agent of state participation in offshore operations. The BNOC began operating in 1976; it took over the North Sea assets of the NCB and those purchased from Burmah Oil. It is intended to be both a center of expertise within the public sector and a commercial oil company. Companies are obliged to enter participation agreements with the BNOC, which gives the BNOC a vote on the operating committees for the fields and an option to purchase oil. This prerogative clearly facilitates a fast and costless transfer of technical knowledge to the state company; it also gives the BNOC a commercial insight into all North Sea operations. There has been much criticism from the oil companies and Conservative politicians opposing the dual role of the BNOC as operator and source of state expertise. The Conservatives are pledged to change the position of the BNOC but are finding it more difficult than they expected to reverse the complex legislation and its commercial ramifications.

In the fifth round of licensing, participation was carried further by making the BNOC the majority shareholder in each license. The BNOC expects, through its own production, through royalties in kind, and through participation agreements, to have access to over one-third of U.K. production in the 1980s.

The fiscal regime prevailing in regard to the North Sea resources was introduced by the Oil Taxation Act 1975. It designated a Petroleum Revenue Tax (PRT) designed to tax the highly profitable fields most heavily, thereby maximizing the state revenue from fields that the oil companies cannot afford *not* to develop, yet not discouraging the development of marginal fields. For fields developed during the first four rounds of licensing, the average government take (from all taxation) is expected to be about 70 percent of net revenues. For fields developed under the fifth round, under which 51 percent of profits go to the BNOC, the state will take about 85 percent.

These expectations were not initially fulfilled. Revenue from the sale of oil in 1980 was about £8.9 billion and of gas £0.6 billion; the government take through PRT and Corporation Tax was £3.8 billion. In part this reflected the initial effects of allowances against taxation, but it was also the only-to-be-expected result of oil-company ingenuity. A supplementary tax was therefore introduced by the Thatcher Administration applicable from January 1, 1981.

Capital investment in 1980 in North Sea oil and gas was £2.4 billion, 6 percent of total capital investment in the United Kingdom that year. Cumulative investment to 1980 is about £21 billion at 1980 prices.

The recoverable reserves in the North Sea are of course uncertain. The present Department of Energy planning estimate is between 2,175 and 4,350 million tons. Production has recently risen steeply, as shown in table 7-1.

The first priority of oil policy is self-sufficiency, to achieve security of supply and to gain the associated tax revenues. The Treasury's desire for rapid access to revenue is offset by the consideration of security and balanced

Table 7-1
British Oil Production in the North Sea, 1976-1980
(millions of tons)

Year	*Production*
1976	11.6
1977	37.3
1978	52.8
1979	76.5
1980	78.7

Source: U.K. Department of Energy, London.

competition. The choice of net self-sufficiency is more arbitrary than it looks. Because of the nature of U.K. refinery capacity, it cannot handle a feedstock made up completely of light North Sea oil; it will be necessary to go on importing Middle Eastern crude oil. A more restrictive depletion policy would be to produce enough light North Sea oil to displace the maximum feasible volume of imports; any figure above that level implies export of North Sea oil.

There are two conflicting considerations which govern depletion policy. On the one hand it is necessary to take measures to control output in the short and medium terms, on the other hand it is important not to inhibit the exploration necessary to find the resources necessary in the long term to prolong self-sufficiency. The powers for the first objective have already been taken; they have since been modified by a number of assurances about the application of the government's depletion policy (the Varley assurances); the general effect is that there will be no cuts in production before 1982 and no delays in development of resources found before 1975.

The general view is that present policies will produce a slight excess of production above net self-sufficiency between 1981 and about 1988; thereafter the United Kingdom will again become a net importer of oil unless further exploration reveals substantial new resources.

The important question at the moment is whether government is providing a suitably encouraging environment for exploration to find those fields, generally smaller and in deeper water, which are necessary to prolong net self-sufficiency. More precisely, questions are:

1. What additional volume of oil must be found to achieve the production goal?
2. What proportion of future discoveries must be commercial to achieve that goal?
3. What level of industry exploration is necessary to find and develop sufficient commercial fields?

The oil industry is lobbying for

1. the government to substantially increase licensing and sustain exploration;
2. fiscal policies to ensure adequate cash flow for exploration and development and a reduction in the threshold of commercial size and development in deeper water; and
3. information and assurances on how the government will use the extensive regulatory controls available to it (oil companies dislike nothing as much as uncertainty).

The high availability of fossil fuels continued to permit an extraordinary lack of decision in choosing nuclear technology. In 1974 the government announced that the Steam Generating Heavy Water Reactor (SGHWR) would be adopted for the next nuclear program. The AGR program had been such a catastrophe that it did not seem possible to seriously propose further construction; the Central Electricity Generating Board (CEGB) as a customer exhibited a variety of opinions but on the whole favored the purchase of U.S. technology and the construction under license of Pressurized-Light Water Reactors (PWRs).

Many elements in the Labor Party found it difficult to accept the idea that the products of a U.K. state agency (the UKAEA) could be inferior to the products of unbridled American capitalism, and were extremely reluctant to take that route. The UKAEA proposed the SGHWR as a reactor that brought together the best elements of Light-Water Reactor and Heavy-Water Reactor technology. This alliance then foisted onto the CEGB a construction program based on the SGHWR. The CEGB were horrified; a committee was established to study the scaling-up of the existing designs to a size suitable for commercial operation. The CEGB did not hide their intention to kill the SGHWR within this committee, and it succeeded. At the beginning of 1978, the UKAEA was forced to recommend to the government that work on the SGHWR be discontinued.

The Electricity Supply Industry and reactor constructors made strong representations to government that unless new orders were placed, soon it would be impossible to keep the industry together and preserve the nuclear option; this, rather than any special immediate need for new capacity, was the driving force behind policy. The only reactor that could be ordered immediately was the AGR, and therefore the CEGB and South of Scotland Electricity Board were authorized to order one AGR each. At the same time the CEGB announced their intention to order a PWR perhaps in 1982, subject to all necessary clearances and consents.

Popular opposition to nuclear power has not so far been strongly expressed in the United Kingdom, and some skill has been shown in keeping opposition out of the courts and out of politics. The origins of anti-nuclear beliefs are complex; countries have developed different modes of handling the conflicts that arise. France has used her traditional ritual of social-conflict resolution—battling in the streets and fields; Germany and the United States have resorted to legislation; in Italy and Spain it seems that regional councils may be used as vehicles for bargaining among local and central priorities. In the United Kingdom the mechanism has been the public inquiry.

A public inquiry has been held on the proposal to extend reprocessing facilities at Windscale, and an inquiry will certainly be held on the PWR and later the Commercial Fast Reactor. Public inquiries were also held on proposals to open coal mines in parts of the country not traditionally mined (Selby and Belvoir). These inquiries permit a great deal of material to be publicly presented and examined; the decision is believed by the objectors to be predetermined; that belief is almost certainly true. But the inquiry does have a significant political effect in separating the hard-core opposition, which is not recoverable anyway, from less-committed objectors; it thereby preempts any mass political movement. Much the same can be said of the Royal Commission report on the Environmental Consequences of Nuclear Power; it displays considerable ambiguity, because of the (successful) attempt to agree in an unanimous report, but the effect is to persuade many of those vacillating on the issue that the problems are honestly acknowledged and have been thought through.

Whether these maneuvers will last is uncertain. There are signs that nuclear power will become an issue with the left wing of the Labor Party (sincerely) and the Liberal Party (less on); the PWR choice will certainly strengthen the anti-nuclear movement; even some members of the nuclear establishment have doubts about aspects of the design.

British policies on energy conservation were set out in a White Paper of July 1976 in which seven roles for government were identified:[12]

1. To ensure that energy prices reflect at least the costs of supply.
2. To publish information about energy saving and to motivate users to save energy.
3. To set an example in the public sector.
4. To ensure the availability of specialist advice and training.
5. To initiate research and development as a means of energy saving.
6. To promote standards and codes of practice relevant to energy conservation.
7. To bring in mandatory measures, where appropriate, but to maintain a voluntary approach on the whole.

A large number of particular measures, which it would be difficult to summarize, have been adopted; it suffices to say that they are consistent with these general principles. The United Kingdom is not a large government spender on conservation, because of a certain scepticism about the effectiveness of resources so employed; this scepticism is even stronger under the Thatcher Administration than previously. Indeed, a serious dissension is now evident between ministers who believe a reliance on market forces to be adequate and senior civil servants who believe that such a policy must be supplemented by governmental intervention. The Department of Energy has studied the idea of an Energy Conservation Agency (à la française) but so far there is no support from the political powers.

Present Issues

The Second Energy Crisis

The United Kingdom is temporarily sheltered from the direct effects of worldwide instabilities in the price and availability of oil. Increases in the price of oil do not cause a large transfer of resources from Great Britain; they do, however, redistribute wealth within the United Kingdom. It follows that the problems experienced after the Iranian revolution are essentially internal; they arise from the allocation of the income created from coal, oil, and gas; the recessive effects of oil revenues on U.K. manufacturing industry; and the changing competitive status among fuels.

The income created by oil-price increases has gone mainly to government and will do so increasingly in the future. North Sea oil prices follow international prices; naturally some of the benefits accrue to the companies but the larger part is extracted through the Petroleum Revenue Tax and BNOC operating profits. In the coal industry, the income appears to have been extracted by the miners in the form of higher wages, but in fact the size of the coal industry and wage levels are determined at the present time by political decision and by prevailing wages for similar occupations; in reality, therefore, the income accrues to government in the form of subsidies avoided.

Gas prices in the nondomestic markets are aligned with oil-product prices and the income has therefore mainly appeared as profits to British Gas (BGC); in the domestic sector prices do not automatically follow oil prices and have been in fact kept artificially low in the interests of political popularity. In this case the major effect of the changes has been to greatly improve the competitive status of gas, and the profit has passed to the domestic consumer. This situation has now been rectified by a decision to raise gas prices by 10 percent per year plus inflation for three years; very large profits will accure to British Gas.

The profits of British Gas and the BNOC are viewed covetously by a Thatcher government committed to reducing or stabilizing the government-borrowing requirement. A statement has been made in principle that a levy will be made on the gas purchased by British Gas from the North Sea and the financial target adjusted accordingly. But such an action would not in fact reduce the overall public-sector borrowing requirement (PSBR) because British Gas is part of the public sector and presently pays part of its profits into the National Loan Fund (£200 million in the last financial year); the only difference would be that British Gas would lose interest on the money and possibly a little financial independence, but as its investments all have to be approved by government anyway, the latter consideration is not substantial. The change only makes sense if nationalized industries were to be taken out of the definition of PSBR; this would be purely cosmetic. Much the same applies to the BNOC—nothing has been decided about the syphoning of BNOC profits, but whatever happens it will be largely a cosmetic change. The real question is what government will do with the revenues that accrue to it directly and from state enterprise.

The present government is opposed to the provision of more extensive state services; it has indicated that the revenues should be used to restructure British manufacturing industry, and in this pious hope it is joined by the opposition. In fact this desire runs quite counter to the economic character of the country and there is no sign of its realization, indeed the reverse is more likely.

The public-sector borrowing requirement being roughly what is would have been without oil, it follows that until now the oil revenues, coming in the first instance to government, have been distributed to consumers as tax reductions. The structure of the U.K. economy is heavily biased toward distribution and services (about 45 percent of total consumption; at the margin it is probably higher.) It will not be possible to import the kind of services on which consumers will want to spend their oil revenues, but manufactured goods can be imported and the natural tendency will be for domestic resources to be switched into services and the domestic manufacturing industry to suffer a depressive effect over and above that arising from the world recession. This syndrome is augmented by the effect of the oil premium on the exchange rate.

The only way of preventing North Sea oil from imposing a contraction in manufacturing is to run a current-account payments surplus by exporting capital, thereby forfeiting immediate consumption; this policy could also ensure that the windfall gain is converted into a permanent source of income, sufficient to sustain a permanent increase in consumption. It is unlikely that overseas investment will be sufficient to offset entirely the recessive effects on manufacturing; indeed to do so would require investing all the revenues and the interest in manufacturing. A shift in the structure of the U.K. economy as a consequence of its oil revenues therefore appears inevitable.[13]

Competition among fuels has been increasing steadily since 1945; energy policy is based on the concept of the United Kingdom as a well-developed four-fuel economy; in reality it is a six-fuel economy because imported coal and oil, although technically similar to the indigenous products, are worlds apart politically. Imports are live policy options; with the existing refining capacity of the United Kingdom it would be extremely costly to use all North Sea oil to replace imports, but there is some scope for modifying the balance between imports and indigenous production. Imported coal has a substantial (20-percent) cost advantage at some U.K. power stations; the CEGB is planning port facilities to handle 10 mt/year (as much as anything this is to put a brake on prices of indigenous coal). The British

Steel Corporation is making strenuous attempts to import coking coal. If conservation is added to the list then there are seven policy options in economic competition.

The option with the competitive edge is gas; and it is the price and availability of gas that is the main source of tension. A recent policy Green Paper by the Department of Energy puts proven gas reserves at 28.6 trillion cubic feet (tncf), probable reserves at 9.6 tncf, and possible reserves at 12.8 tncf.[14] Much higher figures (up to 100 tncf) are known to be contemplated by the BGC. A slightly later note by the department increases its estimate of reserves by some 6 tncf.[15] There is also the possiblility of purchasing further supplies from the Norwegian sector. Much of the expected future supply of gas is expected to be found in association with oil in the Northern Basin, usually in quantities too small to justify the construction of individual pipelines. These supplies can be made available through a gas-gathering pipeline; the important decision is to make the initial construction; subsequent additions are always possible. There is a strong case for bringing forward a U.K. gas-gathering pipeline in order to encourage subsequent Norwegian gas to be landed in Great Britain rather than on the Continent. The latest study of a gas-gathering pipeline, by the BGC and Mobil, has been submitted to the Department of Energy. It proposes an initial pipeline network costing around £3 billion which would be in operation by the mid 1980s, collecting some 400 billion cubic meters of gas.

The finance of this line has been very difficult. Negotiations between the BGC, the oil companies, government, and the banks have hinged on whether the government would guarantee loans for the project. The Treasury has refused to do that on the grounds that such a guarantee would affect the public-sector borrowing requirement. It has refused to allow the BGC to finance the project itself. The oil companies have gone to great lengths to exact concessions on taxation and the BGC's monopsany in return for financial participation. The financial institutions in London regard the economic prospects as uncertain, as they would be controlled under present circumstances by BGC decisions on market penetration, depletion, and price.

At the time of writing it looks as if all participants have gone too far and the gas-gathering system will not be built. This would be one of the costlier consequences of obsession with the public-sector borrowing requirement, which currently wags the tail of the energy-policy dog. This consequence stands wisely in continuity with the nationalistic prejudices of Norway and the United Kingdom, which prevented the construction of the logical central gas-collecting line from fields in the jurisdiction of both countries. If the gas-gathering line does not proceed then the oil companies will probably develop individual selected fields.

The Frigg field is now onstream; with the Brent field this should bring total gas supplies to 6 billion cubic feet per average day (peak capacity 12 billion cubic feet per day) by the mid-1980s; supplies should continue at this level to the 1990s. The BGC proposed to spend £4 billion over the next five years on the necessary facilities. It is possible that synthetic natural gas (SNG) will be the economic substitute thereafter and that by 1990 the technology and primary fuel (almost certainly coal) will need to be chosen. It is more likely that further natural-gas supplies will become available, postponing the advent of SNG.

In order to avoid sudden surges in gas availability as fields come onstream (surges similar to those which in the past have heightened discontent in competing industries), the BGC proposes to use their own fields, Morecombe Bay off the west coast and Rough Field off the east coast, to regulate penetration. The Rough Field will be used as a storage depot by injecting gas at off-peak times; the new field at Morecombe Bay will be used as a peak-lopping device.

Despite these precautions, the increase in natural-gas supplies is having and will have a depressive effect on the electricity industry and therefore indirectly on the coal and nuclear industries; it is also not conducive to energy conservation, which like all other options takes a long time to get going even in a favorable environment.

The consequence for nuclear power deserves some elaboration. System costs for nuclear power of the CEGB system are allegedly negative; that is, fuel savings are supposed to more than justify the cost of installation. But as prices are linked to coal and therefore to oil, electricity as a general commodity is suffering in competition with gas, and there is no apparent justification in terms of demand for system expansion. In other words, marginal costs are below present prices, but demand and therefore investment are determined by present prices. Although increases in oil prices improve the prospects for nuclear power, they depress demand for electricity.

In this situation of temporary excess and perceived long-term deficiency, the obvious solutions are to export part of the surplus or to control depletion of the hydrocarbons. Natural gas is not easily exported, nor is its depletion easily controlled, because so much of future supply is likely to be associated with oil. Oil is the most likely candidate both for export and depletion control.

The United Kingdom in the European Communities (EC)

On the face of it there is considerable benefit to be gained from optimizing energy policy with the EC. Resources allocated to coal-mining are inefficiently directed; vast subsidies are paid to Continental pits despite the existence of much better prospects in Great Britain. The nuclear-construction industry is made up of exclusively national units and must be the most expensive structure that could be devised for absorbing U.S. technology. Kraft-Werk-Union in Germany makes a reactor apparently more reliable than the original U.S. design and with attractive safety features, yet is drastically short of orders; Framatome in France also has extensive under-utilized custom-built manufacturing facilities; yet the United Kingdom and Italy intend to add to this overcapacity by manufacturing their own PWRs under license from Westinghouse. The scope for bargains (for example EC resources to develop coal in the United Kingdom in return for concessions on oil, or common development of LWR technology), is considerable, but the political difficulties are apparently too great for progress to be made.

The United Kingdom argues that any substantial increase in the volume of her oil production would be an unacceptable sacrifice. As for prices of exports, she argues that these must be at world-market levels. It is indeed hardly reasonable to expect Great Britain to offer oil to Europe at a discount. Light oil such as that from the United Kingdom commands a premium, and in general, U.K. oil is sold at a fair price adjusted slightly later than that of its immediate competitors Nigeria and Libya. Disposal of oil in an emergency is governed by IEA rules and therefore no EC preference is possible.

The United Kingdom concedes that it is modestly anti-communataire to refine two-thirds of production domestically. She did attempt to find some EC support for coal in exchange for forgoing the benefits of refining, but no satisfactory agreement could be reached. On the whole there are said to be no more concessions to the EC than are already being made. One should bear in mind that the United Kingdom is already making substantial exports, that the great efforts being put by Great Britain into its coal industry to some extent releases oil for the Continent, and the Communities' offshore industry has had a considerable share of the construction work.

In nuclear-reactor construction, strategic considerations are said to demand that the United Kingdom should have its own capability despite the economic penalties and diversion of resources. In coal-mining, the size of the subsidies to Continental producers is determined by social reasoning, not energy policy; the subsidies are not an indication of the price Continental states are prepared to pay for coal; the EC member countries have no interest in importing moderately high-cost U.K. coal in preference to cheap coal from other continents.

In short, despite the inefficient structures of energy supply that exist in the European Communities, and the substantial misallocations of resources, there appears to be little hope of change. Apparently, important economic benefits cannot be achieved because of political constraints.

Notes

1. I am indebted to a lecture by Robin Grove-White of the Council for the Protection of Rural England for much of the content of this section.

2. See, for example, U.K., Select Committee on Nationalized Industries, *Capital Investment Procedures* (London: House of Commons, 1973-1974), p. 65 which concludes that government had failed to meet its objective of "exercising its control publicly and according to well defined ground rules, without interfering with the management functions of the industries themselves."

3. U.K., National Economic Development Office, *A study of the U.K. Nationalized Industries: Their Role in the Economy and Control in the Future* (London, 1976).

4. Richard Crossman, *The Diaries of a Cabinet Minister* (London: Hamish Hamilton, and Jonathan Cape, 1975).

5. *The Financial and Economic Obligations of the Nationalized Industries*, Command 1337 (London: HMSO, 1961). *Nationalized Industries: A Review of Economic and Financial Objectives*, Command 3437 (London: HMSO, 1967).

6. M.V. Posner, *Fuel Policy* (London: Macmillan, 1973).

7. U.K., NEDO, *A Study of U.K. Nationalized Industries*.

8. *The Nationalized Industries*, Command 7131 (London: HMSO, 1978).

9. Posner, *Fuel Policy*, pp. 192-197.

10. *Fuel Policy*, Cmnd 3438 (1967), p. 74. Reprinted with permission.

11. Lord Alfred Robens, *Ten Year Stint* (London: Cassell, 1972), p. 221.

12. *Energy Conservation*, Command 6575 (London: HNSO, 1976).

13. S. Britton, "Reindustrialization is good for the U.K.," *Financial Times*, July 3, 1980.

14. U.K., Department of Energy, *Energy Policy: a Consultative Document*, Command 7101 (London: HMSO, 1978).

15. U.K., Department of Energy, *Energy Projections 1979*.

8 French Energy Policy

Guy de Carmoy

French energy policy should be analyzed in the broader context of the ambition of France to be one of the leaders in the world economy. French performance has been significant in trade, the country ranking fourth among world exporters after Germany, the United States, and Japan. A compelling reason for export promotion was the necessity for the country to import the bulk of its energy consumption. Industrial development and consumer well-being also required a massive recourse to imports of energy, especially oil.

The first step in the study of the French energy policy is to assess the needs and resources of the country. The second step is to define the major policy goals and to describe the instruments—public, semipublic, and private—which are at the disposal of the authorities for the implementation of these goals. On this basis it will be possible to consider the actual policies pursued in the field of each source of primary energy and in the new but essential field of energy saving. The economic effects of the deterioration in the terms of trade following the successive increases in the price of oil will then be appraised.

Needs and Resources

During the years of rapid postwar economic expansion (1960-1973) French consumption of primary sources of energy doubled from 85.6 million tons of oil equivalent (mtoe) to 176.8 mtoe. Oil consumption increased by a factor of 4.3. During the same period, French production was reduced by one-fifth because of a reduction by one-half of the output of the coal mines (table 8-1). As a result net imports increased by a factor of four. France became the third-largest world importer of oil after Japan and the United States, the third-largest importer of natural gas after the United States and West Germany, and the second-largest importer of coal after Japan. Nuclear-generated electricity was making only a slow start. External energy dependence increased from 38.1 percent to 75.9 percent of total primary-energy consumption.

Because of the long lead time in energy investments, no sweeping change could take place during the short period between the first and second so-called energy crises (1973-1979). Consumption increased slightly. Nuclear-generated electricity production more than doubled but accounted only for

Table 8-1
French Primary-Energy Balance Sheet
(million tons of oil equivalent)

Type of Energy	*1960*	*1973*	*Year* *1979*	*1985*	*1990*
Consumption					
Coal	46.8	30.5	34.5	29.0	33.0
Oil	26.9	116.3	108.5	101.0	68.0
Gas	2.9	15.0	23.0	36.0	42.0
Nuclear	—	3.0	8.5	43.0	73.0
Hydroelectric	9.0	9.8	16.0	14.0	14.0
New energies	—	2.0	3.0	2.0	12.0
Total consumption	85.6	176.8	193.5	225.0	242.0
Production					
Coal	38.9	18.1	13.3	10.0	6.7
Oil	2.2	2.0	2.1	3.0	3.0
Gas	2.9	6.9	7.1	5.0	2.8
Nuclear	—	2.6	8.3	43.0	73.0
Hydroelectric	9.0	10.5	14.7	14.0	14.0
New energies	—	2.0	3.0	6.0	10.0
Total production	53.0	42.1	48.5	81.0	109.5
Net imports					
Coal	7.9	11.1	21.2	20.5	26.3
Oil	24.7	114.4	106.4	93.0	65.0
Gas	—	8.0	15.9	30.5	39.2
Total imports	32.6	133.5	143.5	144.0	130.5
External dependence					
Percentage	38.1	75.9	74.9	64.0	54.8

Source: France, Ministry of Industry, *Les Chiffres Clés de l'Energie (Paris, 1979); Le Bilan Énergétique á l'Horizon 1990*, April 2, 1980.
[a]Projected.

4.5 percent of total domestic primary-energy production. Total energy imports rose from 133 to 143 mtoe but the reduction in oil imports was more than compensated by increases in imports of coal and of natural gas.

As of the end of 1978, French reserves of crude oil were negligible: reserves of natural gas (136 billion cubic meters) amounted to around twelve years of consumption and recoverable reserves of coal (286 mtoe) to only nine years of consumption at the present rate. Lastly, French reserves of uranium (100,000 tons) did not exceed ten years of consumption at the projected 1985 level.

Projections for 1985, overambitious in 1974, had to be scaled down. The figures in table 8-1 are to be considered realistic. Consumption is to increase from 193.5 to 225 mtoe, and domestic production from 48.5 to 81 mtoe, imports remaining at about the same level.

Projections for 1990, recently released, call for an increase in consumption to 242 mtoe, domestic production to rise to 109.5 mtoe, and imports to regress to 130.5 mtoe. The expected increase in domestic production between 1979 and 1990 owes essentially to the implementation of the nuclear program and marginally to new energies, while coal and gas production are on their way down. The import policy calls for a moderate increase in coal imports, a sharp increase in gas imports, and a considerable reduction in oil imports. On these assumptions the aggregate share of coal and gas would remain at 30 percent of total primary-energy requirements, oil would fall from 56 percent to 30 percent and nuclear energy would jump from 4.5 percent to 30 percent (table 8-1). Thus nuclear-generated electricity is the major balancing factor between slowly growing need and dwindling coal and gas domestic-energy resources. It is hoped that external dependence would fall from 74.9 percent in 1979 to 54.8 percent in 1990.

But the French balance sheet should be considered not in isolation but in a European setting. The comparative situation in table 8-2 shows that of the five largest national economies in the European Communities (EC) France ranks third after Germany and Great Britain in terms of total energy requirements and fourth in terms of energy consumption and electricity consumption per capita. These figures reflect the fact that French industrial development is lagging behind that of Germany and has only in recent years approached that of Great Britain. In terms of production of primary-energy sources the output of France barely exceeds one-third of that of coal-rich Germany, one-fourth of that of coal-, oil-, and gas-rich Great Britain, and one-half of that of gas-rich Netherlands. Only Italy is lower than France in this respect. The production situation is reflected in the level of energy imports. The volume of French imports is very close to that of German imports while the French industrial plant is two-thirds the size of the German one. Italy is also a large importer. The Netherlands are self-sufficient and North Sea oil is pushing Britain to self-sufficiency and possibly in a few years time to the position of net exporter (table 8-2).

The industrial development in France was hampered in the nineteenth century and in the first half of the twentieth century by limited domestic-energy resources. By necessity the French were bent on energy saving. In spite of the post-World War II industrialization, the per capita primary-energy consumption and the per capita electricity consumption are below the EC average. By contrast, Great Britain, with a lower per capita GDP (Gross Domestic Product) than France, is much less energy-conscious. Thus France's energy position is not only precarious per se, it is at a disadvantage compared to that of its major Common Market partners, with the exception of Italy.

Tables 8-3 and 8-4 compare the evolution of gasoline and energy prices over time in five EC countries. The general trend has been to favor the auto-

Table 8-2
Energy Economic Features of Selected EC Countries

Energy Use/Production	*Year*	*France*	*Germany (FRG)*	*Italy*	*Netherlands*	*United Kingdom*	*EC*
1. Consumption *(million tons of oil equivalent)*	1978	183.6	258.0	133.5	64.8	208.8	942.3
2. Production *(million tons of oil equivalent)*	1978	40.0	108.9	23.8	71.0	157.8	406.5
3. Net imports *(million tons of oil equivalent)*	1978	143.9	161.2	116.2	3.6	54.3	554.2
4. Percentage of energy from external sources	1978	76.4	58.9	82.4	4.8	25.7	57.0
5. Percentage of primary-energy sources in domestic consumption	1977						
Coal and lignite		16.0	28.9	6.4	4.7	34.2	21.8
Oil		61.1	51.6	67.3	41.5	43.2	53.9
Natural gas		10.2	15.0	16.4	51.7	17.3	17.1
Nuclear energy		2.3	3.2	0.5	1.3	4.7	2.9
Primary electricity		10.3	2.0	9.1	0.3	0.5	4.1
6. Per capita primary-energy consumption *(kilograms of oil equivalent)*	1976	2,935	3,970	2,200	4,170	3,530	3,270
7. Per capita electricity consumption *(kilowatt hours)*	1976	3,875	5,440	2,930	4,200	4,940	4,315

Source: 1, 2, 3, 4, 5: Eurostat, *Basic Statistics 1979*, tables 49, 46, 48, 52, 51.
6, 7: France, Ministry of Industry, *Les Chiffres Clés de l'Energie* (Paris, 1979), pp. 19, 97.

Table 8-3
Evolution of Gasoline Prices in Selected EC Countries
(percentage ratio of retail price index to consumer price index, 1970 = 100)

Year	France	Germany (FRG)	Italy	Netherlands	United Kingdom
1953	117.7	179.2	155.1	113.0	—
1963	123.3	128.7	86.9	104.7	—
1970	100.0	100.0	100.0	100.0	100.0
1973	88.6	92.0	92.5	92.7	86.5
1979	114.0	102.6	114.9	92.9	86.2

Source: *European Economy* 4 (Brussells: EC Commission, November 1980), see table 10.5, Annual Economic Balance 1979-1980.

mobile user. Compared with 1970 prices, the retail price of gasoline has made a small contribution to the reduction in demand. The variation in relation to the consumer-price index was an increase of 14 percent in France and in Italy compared with a decrease in the Netherlands and in Great Britain. It is in France and in Italy and also in the Netherlands that the comprehensive energy-price index has most increased as compared with Great Britain and Germany. Specifically, in 1974 the French industry was sacrificed to the automobile as the price of industrial fuel was increased by 91 percent and the price of gasoline by 29 percent. During the 1973-1979 period, gasoline consumption grew by 19 percent and railroad traffic fell by 6 percent, while it is common knowledge that on medium and long distances rail transportation consumes four or five times less energy than road transportation.[1]

Goals and Policy Instruments

The French government has pursued for half a century a deliberate policy of developing domestic and overseas energy resources. In the interwar period, the major goal was to set up a national oil company and to create a refining industry, both French- and foreign-owned, in France. After World War II,

Table 8-4
Relative Energy Prices for Consumers, 1974 and 1978
(percentage of 1973 price)

Year	France	Germany (FRG)	Italy	Netherlands	United Kingdom
1974	122.1	110.0	125.2	105.7	108.0
1978	121.1	110.4	125.5	121.4	109.6

Source: *European Economy* 4 (November 1980), see table 10.6.

the search for oil was actively pursued in the French overseas territories, a second national oil company was set up, and a long-range program was launched in view of developing a national nuclear industry capable of providing a substantial share of the country's primary-energy needs and of exporting its technology. Since 1974 the policy was geared at speeding up the nuclear program, diversifying the sources of imported energy, and stabilizing energy consumption.

The government has the means to implement these goals through a highly centralized state-control system over the agencies and companies operating in the various sectors of energy. In 1946, the sectors of coal, gas, and electricity were nationalized. The state became therefore the sole owner of Charbonnages de France (CDF), Gaz de France (GDF), and Electricité de France (EDF). CDF has a monopoly and EDF a quasi-monopoly on production in their respective fields. EDF and GDF have a monopoly on transportation and distribution. At the same time the government created the Commissariat á l'Energie Atomique (CEA), a state-owned agency in charge of research and development in both the civilian and military use of nuclear energy. CEA has a 30 percent participation in the capital stock of Framatome, the sole constructor of nuclear boilers. This company is up to 51 percent under the control of Creusot-Loire (of the Franco-Belgian Empain group). The remainder of the capital stock (19 percent), is held by the licensor, Westinghouse, and is to be acquired by CEA in 1982. As CEA has an operational role in the development of the fuel cycle, it concentrated its industrial activities in a wholly-owned subsidiary, COGEMA. Under 1928 legislation, the state holds a monopoly on imports of crude oil and petroleum products, the operation of which is delegated in the form of import quotas to the French and foreign companies and to oil traders under the supervision of the Ministry of Industry. The domestic market is strictly regulated. The quota allocation has tended to slightly favor the national companies. The state has a 35-percent share (with a 40-percent voting right) in the capital stock of Compagnie Française de Pétroles (CFP) and a 70 percent share in the capital stock of Elf-Aquitaine. The foreign companies are Esso, Mobil, Shell, and British Petroleum (BP).

The operating companies, whether partly or fully state-owned, are under the supervision of the Minister of Industry, assisted by the Director General for Energy and Raw Materials. Policy options are debated among the companies and the relevant directorates of the Ministry of Industry (figure 8-1). A representative of the Treasury participates in the debates. The proposals are submitted for decision to the Council of Ministers or more often to a restricted interministerial council (Le Conseil Central de Planification). When major policy decisions are at stake, the interministerial council is presided over by the President of the Republic; the Prime Minister and the Minister of Finance attending, and the Minister of Industry submitting

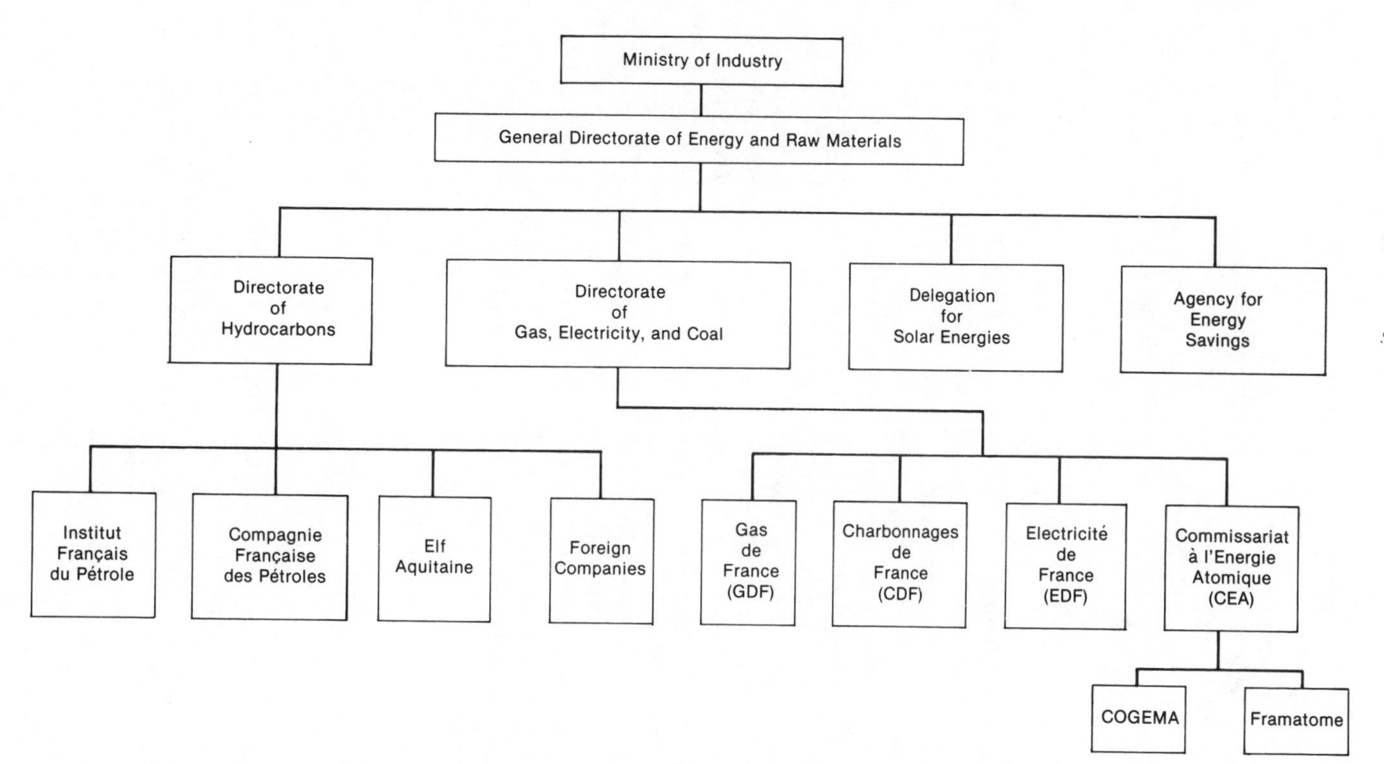

Figure 8-1. The Instruments of French Energy Policy: Public Administrations, Operating Companies, and R&D Institutions

proposals for action. The National Assembly has an opportunity to discuss the energy policy during the annual discussion of the budget and when a new piece of legislation is being proposed. The General Planning Office (Commissariat Général au Plan), whose head attends the interministerial council, incorporates the decisions in its forecasts. The decision-making process is made easier by the common origin of and close relationship among the company managers and the top civil servants.

Coal

French coal has no future except for in a small part of the Lorraine basin. Indigenous production declined from 27 mt in 1973 to 20 mt in 1979, with a further decline to 10 mt foreseen by 1990. The average cost per ton of domestic coal exceeds by one-half the average sales price, and amounts presently to double the cost of imported steam coal.[2]

The ratio of coal imports to coal production will be 3.2 to 1 by 1990. In this circumstance, investing in overseas coal mines is a means of assuring future supplies. To this effect, Charbonages de France has associated with Compagnie Française de Pétroles and with COGEMA, the subsidiary of CEA. A new coal terminal has been built at Le Havre. It can accommodate cargoes of 140,000 tons. The French shipping business is ordering four freighters of that size.

It remains to find outlets for imported coal. The needs of Electricité de France after a peak in 1980-1981 will decrease in relation with the growth of nuclear production. The goal is to promote coal consumption in industry and in collective heating. The conversion in the cement industry is well under way. The conversion of boilers in manufacturers will meet with technical and economic difficulties.

Oil and Gas

The Companies

France had to compensate for the lack of domestic resources through the control of overseas oil fields and therefore through the setting up of strong national oil companies. CFP and Elf-Aquitaine rank third and fifth among the European-based oil companies after Shell and BP, with ENI (the Italian national oil company: Ente Nazionale Idrocarburi) ranking fourth. Their sales in 1978 amounted to 56.3 and 41.1 billion francs respectively.

CFP was the partner of three British and American majors in the Iraq Petroleum Company which was nationalized in 1972. It had a 6 percent share in the Iranian Consortium, which was dissolved in 1979. It lost half of its assets in Algeria. It has shifted to the position of operator with purchasing rights on crude oil, mostly in Algeria, Abu Dhabi, and Indonesia. Elf-Aquitaine is the result of a succession of state-sponsored mergers. It has a solid metropolitan base with the natural-gas field of Lacq in southwestern France. It has developed oil and gas fields in West Africa and especially in Gabon, and has concessions in Canada. Both companies are associated in the Norwegian sector of the North Sea, where they struck mostly gas (Frigg and Ekofisk). Owned resources and purchasing rights account for some 70 million tons per year, an amount substantially lower than the French domestic consumption. The policy of geographical diversification has been extended to downstream activities in a number of industrialized and developing countries. In metropolitan France, thanks to the import-quota system, the two groups have a market share of 53 percent compared with 26 percent for Shell and BP and 17 percent for Esso and Mobil.

In the meantime, l'Institut Français de Pétrole built a network of service contracts and offshore expertise through its engineering subsidiary, Technip. A Franco-American drilling company, Schlumberger, became a world leader in geophysical exploration technologies.

The Suppliers

In 1978, up to 79.1 percent of crude oil was imported from the Middle East, the larger suppliers being Saudi Arabia (35.1 percent), Iraq (17.7 percent), Iran (9.1 percent), and Abu Dhabi (6.4 percent). In Africa, the main supplier was Nigeria (7 percent), Algeria and Libya accounting jointly for 6.1 percent. Efforts are being made to increase the contribution of Iraq so as to compensate for the fall of imports from Iran and to diversify the range of suppliers through state-to-state contracts and trade compensation agreements: Mexico, Venezuela, and Norway are cases in point.

The bulk of natural-gas imports is purchased from the Netherlands (55.6 percent in 1978). Algeria ranks second (19.2 percent) followed by the Soviet Union (14.8 percent) and the Norwegian sector of the North Sea (10.4 percent). The project for deliveries of Iranian gas to Western Europe has been cancelled. Gaz de France is therefore negotiating for increased deliveries of Algerian gas. Nigeria will become a supplier to Western Europe in 1985 according to a protocol signed in February 1980. Cameroun may follow.

Oil Diplomacy

Company expansion abroad and increasing needs for overseas supplies required the full backing of the French diplomacy. It was given unreservedly after the major policy statement made by President de Gaulle at a press conference in November 1967 after the Six-Day War. The President stressed that after the Algerian war France had resumed her ties of friendship and cooperation with the Arab people of the East. This, he added, represented a "basic tenet of our foreign policy."

After the 1973 Arab-Israeli War, and as a result of a French initiative, the EC member states in November adopted an interpretation of the U.N. Resolution No. 242 directing Israel to end its occupation of the Arab lands seized in 1967. France, among other Western European countries, refused to facilitate the flow of American arms to Israel. France, Britain, and Spain, in view of their support of the Arab cause, were declared friendly countries and not subject to the 1973-1974 oil embargo.

When the United States proposed at the February 1974 Washington conference to set up an organization of the industrialized oil-consuming countries, France alone among the EC member states opposed the project on the grounds that these countries should pursue a policy of cooperation and not of confrontation with OPEC. France did not join in the International Energy Agency (IEA) created in 1974, but did not oppose its location in Paris under the aegis of the OECD. This led to procedural difficulties when the IEA and the EC had to discuss issues of common interest, such as emergency energy-sharing and market regulations. President Giscard d'Estaing tried a new line of negotiations by promoting and convening in December 1975, with the lukewarm support of the United States, the Conference on International Economic Cooperation (CIEC), dubbed the North-South Conference. Three groups of countries—the industrialized oil importers, the developing oil consumers, and the oil exporters—discussed not only oil matters, but also raw materials and development. The conference collapsed in July 1976.

Meanwhile, under French initiative, the EC member states attempted to engage in a so-called Euro-Arab dialogue with the Arab League. At the request of the United States, the subject of oil was removed from the meetings that took place in 1974. The political issue of representation of the Palestinian Liberation Organization (PLO) troubled the dialogue, which did not become an international force.

France nonetheless pursued its pro-Arab policy, which gained momentum in 1980 during President Giscard d'Estaing's tour in the Persian Gulf countries. His statement on the right of self-determination of the Palestinians had some impact considering the time and place of delivery. It was not criticized by the EC member states.

In its relations with the EC, France was induced to apply greater flexibility to its import monopoly on crude oil and petroleum products. The quota system was ill-adapted in a situation of overcapacity in refining. Quotas were suppressed in 1979 but import authorizations were maintained on the basis of three-year import programs presented by the operators. Furthermore, the importers were authorized to purchase oil within limits on the free market. Thus the domestic market was becoming more open to competition.

The Oil Industry

Consumption in France and in Europe leveled off after 1973, while the oil companies had increased their tanker fleets and their refining capacity in the early 1970s in the anticipation of a continuous growth in consumption. As a result, the rates of freight and refining fell barely above the marginal costs. The French companies, short of crude oil outside of the OPEC countries, were at a disadvantage vis-à-vis some of the larger international groups that could compensate their losses in refining with profits in production. The French refineries were working at 64.3 percent of their capacity in 1975 and at an average of 70 percent in 1976-1978. The two companies had to build converter units to adjust to the declining demand for heavy-oil as compared to light-oil products. After a series of lean years, the profitability of both companies increased, as the recovery was faster—and therefore the investment capacity higher—for Elf-Aquitaine than for CFP.

The companies were induced, through subsidies from the 1980 budget, to step up oil and gas exploration and production onshore (in the Paris basin, in Aquitaine, and in northern France) and offshore (in the Western Approaches, in the Aquitaine Gulf, and in the Lion Gulf). Exploration is financed jointly by the companies' cash flow and by a special fund, the Hydrocarbons Support Fund, the resources of which have been increased in the 1980 budget and are scheduled for further increase in 1981. Somewhat contradictorily, an exceptional tax was to be levied on the depletion allowances of the oil companies.

Electricity and Nuclear Energy

Electricity production was generated in 1978 up to 55.5 percent in thermal-power plants, hydroelectricity accounting for 31.2 percent and nuclear electricity for 13.3 percent.[3] Since 1973 the amount of coal in thermal-power plants was doubled so as to spare oil and gas for other uses. Coal-fired power plants are being built accordingly. A survey was made of the country's hydraulic resources to find out if the equipment of new sites would be profitable at current energy prices. Several projects were launched in 1978 that will add about 1 percent to the present production of electricity, but the project for the construction of a large tidal-power plant at the Chausey Islands was shelved. The major technical and financial efforts were directed toward nuclear-generated electricity.

Nuclear Reactors

The first reactors used on a commercial basis beginning in 1963 were of the natural-uranium graphite-gas type. Their operation was satisfactory. They were not competitive with American Light-Water Reactors (LWRs) using enriched uranium. In 1969 after a protracted controversy between CEA and EDF, the government opted for the Light-Water Reactor advocated by EDF. The total installed capacity at the end of 1979 was of 12.0 GWe, of which graphite-gas reactors accounted for 2.3 GWe.

In the field of advanced reactors, CEA concentrated its research activities on breeders. Phenix, an experimental reactor with a capacity of 250 MWe, became operational in 1974. In 1976, the government started to build a first commercial-size reactor of 1,200 MWe called Superphenix. Coupling to the grid is scheduled for 1983. A company, NERSA, was created in which EDF joined forces with its German and Italian counterparts, RWE and ENEL. RWE has reassigned its share to a subsidiary, SBK, which is Dutch. Belgium and British interests have a small participation. The construction is entrusted to a group of French and Italian manufacturers. A symmetrical operation is to take place later for the development of a German-type breeder. France and Germany are pooling their research for breeders and will transfer their technologies to a sole licensor.

Fuel Cycle

COGEMA, the operational branch of CEA, is a vast industrial complex covering prospection, mining and enrichment of uranium, production of fuel elements, and reprocessing.

France has domestic recoverable uranium reserves presently estimated at 100,000 tons, following the opening of a new mine in the south of the country. Annual production of 2,000 tons in 1978 was supplemented by 2,500 tons from mines in Niger and Gabon in which COGEMA holds sizable participation. Resources are also being developed in Saskatchewan (Canada). Thus COGEMA is hoping to dispose of a production of 10,000 tons per annum by 1985.[4] Domestic resources are used sparingly and tend to be regarded as strategic stock.

France had set as a major policy objective the construction of a large enrichment facility so as to dispense with imports of enriched uranium from the United States and the Soviet Union, and possibly to become an exporter in conjunction with the sale of French-built PWR reactors. The EURODIF project is a cooperative venture launched in 1972 between France (with a 52.6 percent share), Italy (25.2 percent), Belgium and Spain (11.1 percent each). In 1973, Iran purchased a 10 percent share through a Franco-Iranian subsidiary. This participation was frozen in 1979 as Iran had given up its nuclear program and cancelled its orders for French-built reactors.[5] Work was started at Tricastin in the Rhone Valley in 1974. A first enrichment plant element became operational early in 1979 and it is expected that the plant will reach its full capacity by the end of 1981.[6]

EURODIF negotiated long-term contracts with its shareholders and also with Germany, Switzerland, and Japan. It will supply fuels for the reactors to be exported by France to third countries. A second cooperative reprocessing project, by the name of COREDIF, postponed because of the worldwide slowdown of the nuclear industry, could be reactivated if the demand justified it. Therefore the industrial development of centrifuge technology, cosponsored by Britain, Germany, and the Netherlands, may not take place. COGEMA might contribute to the construction of a diffusion-enrichment plant in Australia.

The reprocessing of nuclear fuels is taking place at the La Hague plant near Cherbourg. Its present capacity is 400 tons/year (t/y). It will be brought to 800 t/y in 1984, 1,600 t/y in 1985, and later to 2,400 t/y. COGEMA has signed contracts with Germany, Japan, Sweden, Belgium, Switzerland, and the Netherlands for some 6,000 tons of fuel to be reprocessed between the mid-1980s and the mid-1990s. The enlargement of the British plant of Windscale has been authorized in 1979 and the operations are scheduled to start in 1987. Because of political opposition, the construction of a plant at Gorleben in Germany is at a standstill. The consruction of an American plant at Barnwell, South Carolina, with a capacity of 1,500 t/y, was frozen by President Carter's instruction. As a result, La Hague is presently the only plant available for retreatment in the OECD area.[7] COGEMA was able to impose stiff conditions to its clients, obliging them to prefinance a large part of the costs of the plant. France, Great Britain, and Germany associated in United Reprocessors to avoid investment duplication. France has sold a pilot reprocessing plant to Japan that is under construction.

In 1978 COGEMA started the operation in Marcoule of the first plant for the solidification of liquid nuclear wastes.[8] A license agreement has been sold to Germany. After vitrification, the wastes are air-cooled in concrete wells where they are to stay for several decades. The ultimate disposal is to take place in dry geological structures. The choice of the structure is still undecided: salt mines (presently the German option), derelict tunnels, granitic or argillaceous geologic structures. Vitrified waste is not cumbersome: two cubic meters per reactor and per year. It is an inert matter, nonreactive with air and water.

Nuclear Program

France's main reaction to the 1973 oil crisis was to step up its nuclear program. The 1974 goal of 55 GWe had to be scaled down for two main reasons: the capacity of the industry to build the reactors, and the possibility for EDF to secure sites in the face of local opposition. Procedure and site preparation require an average of three years and the construction of a reactor six years.

By the end of 1979, the installed nuclear capacity was 12.0 GWe, including the graphite-gas reactors of the 1960s, the first six PWRs, and the experimental breeder Phenix. By the same time thirty-one reactors were under construction or on order, including Superphenix, which would bring the installed capacity to 39 GWe by 1985. Thus, the share of nuclear-generated electricity in total primary-energy consumption would be raised from 3.4 percent in 1978 to 19.1 percent (43 mtoe) in 1985.[9] Orders by EDF to Framatome were grouped by technical level in three batches so as to benefit from standardized production. The initial plant dimension was 900 MWe. The first order for a 1,300 MWe plant was placed in 1978. Annual orders are in the 4,500 to 5,000 MWe range, according to the size of the reactors. Framatome is equipped to construct two supplementary reactors for export. It has so far sold two reactors to Belgium and two to South Africa; the orders from Iran were cancelled. Negotiations are under way with China. Once implemented, the program would double the share of primary electricity (hydroelectric and nuclear) in the French-energy balance sheet, from 12.3 percent in 1978 to 25.3 percent in 1985.

The long-range program to the year 2000 is more difficult to project. The graphite-gas reactors will be dismantled in the 1990s. The breeders, in a slow-growth assumption, would account for about 20 percent of a total installed capacity of 86 GWe by the year 2000 (EDF is asking for the authorization to launch two 1,500-MWe breeders before 1985). By that time nuclear-generated electricity would represent some 80 percent of total electric power compared with 55 percent in 1985.[10]

Evaluation of the Nuclear Program

The French program should be assessed from the points of view of competitiveness, technical options, and public acceptance.

As of 1978, nuclear-generated electricity appears competitive vis-à-vis other sources of electricity. Table 8.5 compares the production costs in cents per kilowatt-hour (kWh) for coal-fired, oil-fired, and nuclear power plants. Operating costs for nuclear reactors include the transport of irradiated fuels, reprocessing, and waste disposal over several decades.

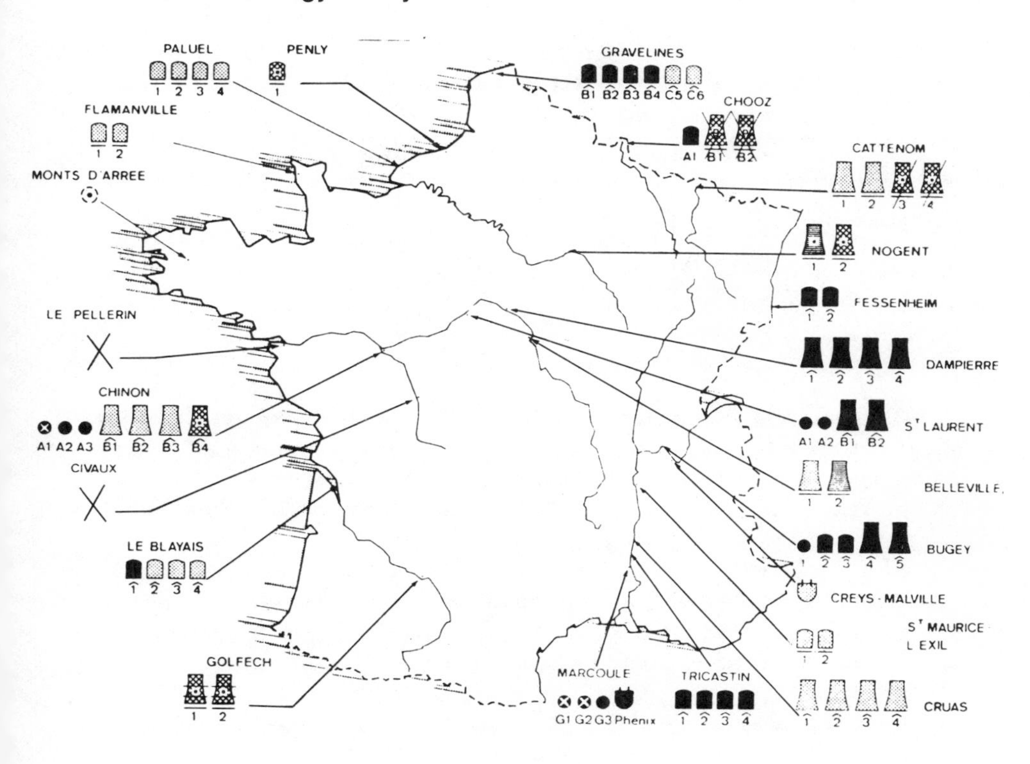

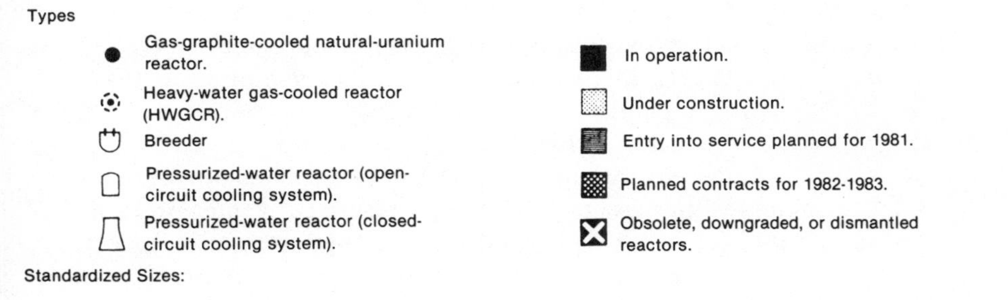

Decision of July 30, 1981, while waiting for the parliamentary debate regarding the future of nuclear energy in France, scheduled for October 1981.

X No work ongoing at site.

/ Work postponed.

— Preliminary work on site.

Map distributed by the Division Information sur l'Energie 3, rue de Messine 75384 Paris CEDEX 08.

Figure 8-2. Nuclear Reactors in France, August 1981

Table 8-5
Comparative Costs of French Electricity Production
(cents per kwh, French francs January 1, 1979)

Costs	*Nuclear-powered*	*Coal-fired*	*Oil-fired*
Investments	6.1	4.7	4.0
Operating costs	2.3	2.6	2.3
Fuels	4.7	16.0	23.7
Total costs	13.1	23.3	30.0

Source: France, Ministry of Industry, *Note d'Information* 31 (Paris: March 1980).

The comparative advantage of nuclear-generated electricity in LWRs should increase in relation with the sharp 1979-1980 increase in oil prices, and in spite of the fact that construction costs of nuclear power plants are on the rise, because of more stringent safety constraints, among other things.

With the breeder reactor, fuel costs would be much lower and investment costs higher. As a prototype (though of commercial size), Superphenix will be expensive—about twice the cost of a serialized PWR of the same installed capacity.[11] No estimate has been made of the costs of a serialized breeder.

France's major technical option was to switch from graphite-gas reactors to LWRs. This was unquestionably a sound though belated option in regard to Germany. The second option in the field of advanced reactors was in favor of the breeder and against the High-Temperature Reactor (HTR). It may well be that the French scientific and industrial community could not pursue both paths at the same time as did Germany.

One should be aware that in renouncing HTRs France jeopardized the prospects for coal gasification and for the launching of a hydrogen industry. The option for the gaseous diffusion process in enrichment does not seem open to criticism. As time goes on, the centrifuge alternative appears more costly and probably more dangerous from the point of view of nuclear-arms proliferation.

A debate is going on concerning the opportunity cost of cogeneration, that is, the simultaneous production and distribution of electricity and heat—either for industrial use or for urban heating. An alternative would be the building of small reactors exclusively for heating. EDF objected both to the cost of small nuclear plants and to their location in the outskirts of the large towns. Proponents put forward various examples in Sweden, Denmark, and Germany. A draft legislation enabling the local authorities to engage in urban heating is presently under discussion in the French National Assembly.

Safety regulations follow the norms set forth by the International Committee for the Protection against Ionizer Radiations. Since the PWR plants are built under Westinghouse license, the regulations are largely identical to those applied to American PWRs, subject to adaptation required by alterations introduced in the French version of the reactors.

The attitude of the general public toward the implementation of the nuclear program has been less critical than in a number of European countries such as Germany, Switzerland, Austria, Sweden, and the Netherlands. A serious riot took place in 1977 on the site earmarked for Superphenix. Trade unionists and local authorities joined with ecological groups in 1978 to oppose the transportation of nuclear waste shipped from Japan to the reprocessing plant in La Hague. Both events had no serious consequences. Strong anti-nuclear demonstrations erupted in February 1980 at Plogoff in Brittany. New elaborate procedures were set up in 1976 for the consultation of the population concerned in the siting of nuclear power plants. The regional council in the regions and the general council in the departments, both elected bodies, are to be consulted on siting nuclear projects. In the case of Brittany, these bodies were invited to choose among five sites. They chose Plogoff, where the opposition came from the municipal council.

Three political parties out of four, the Gaullists, the Giscardists, and the Communists, back the nuclear program. The Socialists, in their majority, favor a national referendum, a slowdown of the nuclear program, and a greater diversification on coal; they are rather hostile to the breeder. Trade unions support the nuclear policy, with the exception of CFDT, a union connected with the Socialist party. A recent opinion poll revealed that a majority of 60 percent favored the program, contrasting with a 1977 poll resulting in an opposition of 60 percent. This change is comparable to that which took place in Sweden. At the level of the elites there is a growing awareness of the precariousness of the country's energy situation. Lastly the highly centralized administrative system under the authority of the Prefect in the department plays into the hands of the government, whose goals never wavered in this specific field.

International Relations

France is engaged in a number of cooperative ventures with several European countries regarding various stages of the fuel cycle and breeder development. The cooperation has been much closer with the Continental countries than with Great Britain. Like the other EC member states, France is a member of Euratom, whose goal was to promote the nuclear industry on a European basis. The goal was not fulfilled because national policies prevailed. Because the supply monopoly on uranium devolved to Euratom does not correspond to present-day economic realities, France has asked her partners to amend the Euratom treaty to remove this discrepancy. The question is under consideration.

Of much more serious concern to the French government is the American nuclear policy. The deferment by President Carter of nuclear-fuel reprocesing and of the construction of the breeder reactor means that if a number of Light-Water Reactors became operational, then the United States might become a net importer of natural uranium, thus endangering European supplies. On the international front, the United States is eager to prevent nuclear proliferation. In 1977 it convened a conference in London on International Nuclear Fuel Cycle Evaluation (INFCE). The conclusions of this conference go against some Amercian theses, and justify the breeder not only as a means of increasing energy efficiency but also from the point of view of safety, as it produces fewer radioactive releases, less thermal pollution, and less nuclear waste.[12] At the same time, a CEA research team worked out a low-grade enrichment process for uranium that would make it improper for developing atomic weapons. France and the United States signed an accord to evaluate the French system over a one-year period and to study the possibility of building a pilot plant using such fuel.[13]

Investments and Tariffs

Massive investments are required for the extension of oil prospection and production, for the development of the nuclear program, and for the appropriate energy substitutions. In the Seventh Plan (1976-1980), energy investments were estimated at 9.5 percent of gross domestic capital formation. This proportion might be short of the needs in the coming years.

The share of self-financing and the borrowing capacity of the operators are dependent on their actual profits, and profits are dependent to a large extent on tariffs, which are in the hands of the government. Generally speaking, tariffs have not been strictly related to costs and operators have been registering large deficits. The volume of cash flow and of investments in the oil sector remained at the same level in current francs and therefore decreased in real terms between 1973 and 1977.[14] The indebtedness of the companies engaged in oil refining and petroleum-products distribution reached 57 percent of pretax turnover in 1977.[15] The investments by EDF in 1978 amounted to 20.2 billion francs, of which only 39.6 percent was met by self-financing.[16] Investments will remain at the same level in constant francs throughout the 1980s, but the profit-and-loss account should be balanced by 1982.[17] The total debt of EDF at the end of 1978 reached 66 billion francs, of which some 13 billion was borrowed on the international market. With the December 1979 rise in OPEC oil prices and the need for EDF to maintain its level of investments in real terms through the 1980s, the financial situation of the enterprise was becoming critical. In January 1980, the government decided to write off EDF's debt to the state of 11.7 billion francs and to increase by an equal amount its capital, so as to improve its credit worthiness.[18] In January 1980 the decision was made to immediately adjust all tariffs to the increases in the prices of imported energy.

Energy Saving and New Energy Sources

The purpose of energy saving is to reduce energy consumption without hampering economic development. Action was engaged at three levels: fighting against waste of energy through a change in the behavior of the consumer; changing the equipment through appropriate investments; and promoting R&D of new technology. Agreements were entered into with trade organizations after a survey of the operations of the main industrial energy users. Thermal isolation of new buildings was encouraged. Information on the energy consumption of automobiles and appliances was made compulsory. The specific goal was to gradually reduce consumption up to 45 mtoe/year by 1985 so that the elasticity ratio of consumption to income would be brought down from 1 to 0.8 in 1976-1980 and to 0.5 in 1981-1985 (for 1979 the elasticity ratio was 0.77). Actual savings amounted to an average of 13 mtoe during the 1975-1977 period, thus reducing the rate of external dependence by about 1 percent. Under the regulations in force in 1979, the risk was that energy saving would not exceed 25 mtoe by 1985. The energy-saving policy will not work as long as the government does not correct serious mistakes such as subsidies to consumption through prices.[19] In April 1980 a reinforcement of the incentives for energy saving in the three sectors of industry, housing, and transportation was decided upon.

It is also expected that appropriation for research on new energy sources will increase. By the year 2000 the entire geothermal resources will be in use in some 800,000 housing units. By the same year, 6 million housing units will be equipped with solar-generated hot-water systems. Solar space heating requires a much more elaborate and capital-intensive technology and will develop slowly over the next two decades. CEA is engaged in fundamental research on materials for the photovoltaic process.[20] Taking into account the 1980 decisions to accelerate the promotion of new sources and especially of biomass, it is hoped that new sources and technologies will contribute 5 percent of the primary energy requirement as early as 1990.[21]

Economic Impact

The rise in energy prices in 1979-1980 will have much more serious consequences than the rise of prices in 1973-1974. During the 1973-1978 period, the terms of trade deteriorated (on the basis of 1970 = 100 percent) from 103.9 percent to 96.4 percent, and the balance of current accounts ran into a deficit in the order of 1 percent of GDP per year on an average.[22] The proportion of energy imports to total imports increased from 12.4 percent in 1973 to 19.5 percent in 1978.[23] By 1978 the balance of current accounts was back to equilibrium. Internally, the economy had been able to absorb the first oil shock, thanks to a reduction in the rate of investment combined with a low increase in the rate of consumption. Externally, the share of exports in the GDP increased from 17.2 percent in 1973 to 20.0 percent in 1978. Thus, the transfer of wealth to the OPEC countries had taken place through an internationalization of the economy.

Under the impact of the second oil crisis, the current balance is shifting from a surplus of $1.5 billion in 1979 to an estimated deficit of $4.0 billion in 1980.[24] Furthermore, the nonenergy trade surplus is shrinking at the time it should increase so as to compensate for the oil price hike. The competitivity of French prices in manufactures seems to weaken.[25]

Appraisal

As a country poor in energy resources, France should combine oil and gas prospection, nuclear-electricity promotion, and stringent energy conservation. On the domestic front, France has a mixed record of remarkable achievements and serious shortcomings. It was able over the last fifty years to launch two efficient, integrated, and multinational oil companies. It was able over the last three decades to develop the strongest nuclear industry in Europe in terms of mastery of the fuel cycle, of steady flow of LWR construction, and of breeder-reactor experimentation. It enjoys a strong marketing position for the export of products and services related to nuclear-generated electricity. In 1985 France's installed nuclear capacity will be double that of Germany and nuclear-generated electricity will supply more than one-half of its electricity consumption. By constrast, the government has followed until 1980 a lax taxation and pricing policy at the consumer level (though it might be termed a strong policy in comparison with that of other European countries). Energy saving on the basis of the regulations in force in 1979 will not meet the target set for 1985.

On the international front, dependence on foreign supplies is expected by 1990 to fall from 75 percent to around 55 percent of total primary-energy consumption. But oil imports would still be around 70 million tons a year. This forecast rests on the optimisitc but unlikely assumption that OPEC supplies to the OECD countries will remain at about the same level in the next five years.

Under the shock of the December 1979 oil-price increase, the government in April 1980 came to the decision that it must sharply reduce the country's oil consumption over the next decade. The targets for 1990 are drastic: the share of oil should fall from 56 percent in 1979 to 30 percent of total primary-energy consumption; the share of nuclear energy should be raised to 30 percent; and a greater emphasis put on the expansion of the domestic market for coal, on new energies, and on energy saving.[26] These goals are sound but overdue. They require heavy investments. Among the larger European countries, France with its scarce energy resources is with Italy the most exposed to an oil squeeze. On the other hand it cannot afford a currency depreciation which would mean a proportional increase in the oil-import bill. Its industrial specialization and export performance do not compare with those of Germany. Its influence on the geopolitics of oil is handicapped by its absence from the IEA debates and by the lack of a firm EC energy policy. From whatever angle it is considered, over the next decades energy will be the major concern of any French government.

Epilogue: The Energy Policy of the French Socialist Government

Energy policy played a part in the campaign for the 1981 presidential election. Outgoing candidate Valéry Giscard d'Estaing stressed that the implementation of an ambitious nuclear program was one of his foremost achievements during his term of office. Contender François Mitterrand, in the television debate that took place on May 5, 1981 with Giscard advocated a more diversified energy policy, with a larger share for coal, energy conservation, and new energy technologies: "I do not reject nuclear energy. I intend to master it." The ecologists, whose candidate polled 3.87 percent of the vote in the first run, generally backed the Socialist candidate in the second run.

On July 30, the Council of Ministers announced the stoppage of work and procedures on five nuclear-reactor sites and the completion of five others. The decision to interrupt the procedures at Plogoff were made just after the election, in accordance with the promise made by Mr. Mitterrand during the campaign.

According to the plan of the preceding government, the share of nuclear-generated electricity was to increase from 6.7 percent in 1980 to 30.2 percent in 1990, a share equivalent to that of oil. The socialist government appointed after the June 1981 legislative elections proceded to revise the energy program. As a matter of fact, a debate took place inside the Socialist Party, the three other political parties already having supported the nuclear option. The debate centered on three questions: the pace of construction of new power plants, the extension of the nuclear fuels reprocessing plant in La Hague near Cherbourg, and the future of the breeder. Of the five reactor sites suspended in July, three were reactivated in November after consultation with the regional and local authorities. The power plants under construction will be completed. The orders for 1982 and 1983 have been reduced from nine to six PWR reactors. The reprocessing plant will be enlarged in order to implement the reprocessing contracts signed with foreign operators. Superphenix will be completed on schedule, but the decision to proceed with the construction of a series of commercial breeders was withheld at the end of 1981. According to the energy program approved by the National Assembly in October 1981, the share of nuclear electricity in total primary energy consumption would stand by 1990 at around 27 percent with a net power of 56 GW.

As for coal, national production will be maintained at around 20 mt subject to a ceiling put on a government subsidy. Appropriations for energy saving and for research and development of new energies have been sizably increased.

The energy program of the new government marks a slight inflection, not a basic change, from that of its predecessor. In spite of the strong advance in the nuclear field, the French energy position remains fragile. The energy bill amounted to 23.2 percent of total imports in 1980. The imports of primary energy are payable in dollars or other hard currencies. The extent of the burden of these imports depends upon the strength of the franc, and therefore upon the success of overall economic policy.

Notes

1. Alfred Sauvy, "La corde au cou," *Le Monde,* January 13, 1980.
2. France, Ministry of Industry, *Les chiffres clés de l'énergie*)Paris: 1979), p. 89.
3. Ibid., p. 107.
4. "La Cogema fait des bénéfices," *Le Monde,* July 6, 1979.
5. "La participation (10%) de l'Iran dans Eurodif est 'gelée'," *Le Monde,* December 4, 1979.
6. "La Cogema . . .," *Le Monde,* July 6, 1979.
7. J.F. Augereau, "Le retraitement civil dans le monde," *Le Monde,* May 18, 1979.
8. Xavier Weeger, "La premiére installation industrielle de vitrification des produits radio-actifs fonctionne depuis six mois à Marcoule," *Le Monde,* December 27, 1979.
9. France, Ministry of Industry, *Lettre d'Information du Ministère de l'Industrie,* Letter 101 (March 18, 1980).
10. CEA (Commissariat à l'Energie Atomique), *Notes d'Information* 7 and 8, (July-August 1978), p. 6.

11. "Journée d'Etudes sur les surrégénérateurs," *Revue de l'Energie* (February 1980), p. 75.

12. Milton R. Benjamin, "Study Sees Safety Advantages in Nuclear Breeder Reactors," *International Herald Tribune,* February 27, 1980.

13. "U.S. and France Sign an Atomic Fuel Accord," *International Herald Tribune,* September 19, 1979.

14. André Demargne, "Les perspectives énergétiques," Report of the Economic Council, (February 28, 1979; *Journal Officiel,* June 28, 1979, table 26, p. 933).

15. Ibid. (June 28, 1979), p. 960.

16. "Electricité de France, 1978," *Revue de l'Energie* 318 (October 1979), p. 877.

17. "Le Financement à Moyen Terme du Programme Electro-Nucléaire," Lettre 101, Information Letter of the Ministry of Industry, Paris: January 8, 1981.

18. "French Gas Upped: EDF Debts Cut," *International Herald Tribune,* January 4, 1980; Demargne, p. 956.

19. Demargne, pp. 910 and 914.

20. Henry Durand, "Qu'attendre des énergies nouvelles," *Bulletin Acadi* 235 (December 1978), p. 513.

21. France, Ministry of Industry, "Le bilan énergétique à l'horizon 1990," April 2, 1980.

22. *European Economy* 4, (December 1979), Appendix, tables 20 and 29.

23. EUROSTAT, *Basic Statistics of the Community* (1973-1974, table 68; and 1979, table 105).

24. OECD, *Economic Survey No. 27* (July 1980), p. 90.

25. "Le second choc pétrolier et les échanges internationaux, Un déficit plus durable qu'en 1973," *Le Monde,* July 8, 1980.

9 West German Energy Policy

Dieter Schmitt

The energy policy system in the Federal Republic of Germany (FRG; West Germany) can be considered quite liberal compared with that of other European countries. But an analysis shows that German energy policy too is characterized by many more-or-less uncoordinated interventions into the market. Most of these measures have in the past, however, been concentrated on protecting a rather small and diminishing section of the energy sector—domestic coal production. For more than two decades, energy policy in the FRG has stressed the adjustment of noncompetitive indigenous hard-coal production to the conditions of an open (in principle) market. This market has been characterized by low, (and in real terms during the 1960s), decreasing prices resulting from abundant oil supplies, intense competition, and economies of scale. The resulting advantageous energy-supply conditions surely have been one of the most important factors in economic development and increasing welfare in our economy.

The first and second oil crises have totally changed the situation. Earlier we had abundant oil supplies, hard competition, a steady flow of new reserves, new oil provinces, new kinds of energy, technical progress, and buyers' markets with low prices. Now we have world oil prices at a level more than seventeen times higher than that in the early 1970s (and which will perhaps double again within the next decade); and a cartel that controls oil-supply conditions. OPEC's production policy has until now limited output to a level that enabled it to charge according to the anticipated (increasing) production cost of future alternatives—all to the detriment of the world economic system.

Until now, the FRG seems to have suffered comparatively little from the developments on the world-energy markets. Despite increasing exhortation from the government, members of all the political parties, and scientific institutions, to conserve energy; despite more and more energy-saving measures from federal and state governments; despite information campaigns, subsidies, tax credits and regulations; despite a new oil-price increase in 1978-1979 which more than doubled prices for imported crude oil; primary-energy requirements (PER) in the FRG reached a new peak in 1979. With a growth rate of almost 5.0 percent, the PER level for the first time exceeded 400 million tons of coal equivalent (mtce) and reached 408 mtce (see table 9-1). Oil consumption increased too, but not as quickly, by 1.3 percent to 206 mtce, with the share of oil consumption as a percentage of PER still

Table 9-1
Primary-Energy Requirements in the Federal Republic of Germany, 1950-2000
(million tons of coil equivalent)

	Year										
Type of Energy	*1950*	*1955*	*1960*	*1970*	*1973*	*1975*	*1977*	*1978*	*1979*	*2000*[a]	*2000*[b]
Hard coal	98.7	131.5	128.3	96.8	84.2	66.5	67.0	69.2	75.8	107	105
Lignite	20.7	27.3	29.2	30.6	33.1	34.4	35.1	35.9	38.1	45	45
Oil	6.3	15.5	44.4	178.9	208.9	181.0	193.9	203.3	206.8	147	115
Natural gas	.1	.6	1.1	18.5	38.5	49.2	55.5	60.4	66.0	90	80
Imports of electricity	6.2	6.1	6.6	8.4	8.2	7.8	7.3	6.6	5.8	13	10
Nuclear	—	—	0.0	2.1	3.9	7.1	11.8	11.8	13.9	128	100
Other	3.5	2.4	1.9	1.5	1.7	1.7	1.7	1.8	1.8	30	20
Totals	135.5	183.4	211.5	336.8	378.5	347.7	372.3	389.0	408.2	560	475

Source: Arbeitsgemeinschaft Energiebilanzen.
[a]Updated energy program from 1977.
[b]Own estimates.

at over 50 percent. Thus, this quick glance at the energy situation in the FRG through 1979 might provoke the question: What must happen to cause a change in traditional energy trends?

But appearances of no change are deceptive. A more detailed analysis shows, that reactions to the dramatic changes in the world-energy market have commenced and are accelerating. The problems, however, have apparently increased too. Balance-of-payments problems, unknown to the German economy for fifteen years, suddenly have returned to public debate. The amount paid for oil by the FRG increased in 1979 by 15 billion deutsche marks (DM) to almost 50 billion DM, in 1980 to 65 billion DM, and in 1981 may increase to almost 75 billion DM. As a consequence, the German balance of payments in 1979 for the first time since 1965 became negative (by 9.5 billion DM). The deficit in 1980 increased to 29 billion DM. The impact of the recent oil-price increases (and their effect on other energy prices) on the inflation rate in 1980 is estimated to range betwen 0.5 percent and 0.75 percent. It is true that the FRG holds large monetary reserves; that balance-of-payments difficulties as well as inflation have many sources; and that energy policy is hardly the adequate instrument to solve these macroeconomic problems. But the influence of the energy sector upon them is undeniable.

Because of the increasing diversion of private capital to pay for expanding energy bills (expenses that have not been balanced by purchases by energy-exporting countries), there has been an overall negative effect on both employment and economic growth. Investment decisions have also been badly affected by the negative expectations of industry. Needless to say, these are serious problems for an export-intensive country like the FRG, and only in the long run do we expect our economy to benefit from the adjustment process of finding substitutes for oil, a process which will ultimately change our industrial structure and energy-consumption patterns.

What answers are there within the energy policy of the FRG to these problems? Is it justified to allow prices to determine the market? Do the market forces adjust quickly enough to new situations? Are the traditional instruments suitable and sufficient? Do we need a more active role for government? Is a totally new course necessary for energy policy? This chapter tries to describe West German adjustment efforts and their preliminary results for the FRG, and the role that energy policy will play in this process.

We must start with a fairly comprehensive overview of German energy policy in the three decades since World War II. This is essential because the fundamental approach of current energy policy was developed in the 1950s to 1970s; because past energy-policy measures will probably remain in force in the future; and because the level and structure of demand and supply today, as well as in the near future, have been largely determined by decisions made (or not made) in the past.

The post-1973 changes in the world-energy situation occurred just as the FRG published a major review of its energy program (updated in 1974, and 1977).[1] On the basis of an analysis of supply and demand and an evaluation of the given energy-policy options, the German government described in these programs its views about its actual and expected energy problems, and measures and instruments to cope with these problems. Basic insights included the view that despite far-reaching conservation in the next two decades, energy demand will increase by about 20 percent. This increased demand will have to be supplied from the world market because the possibility of increasing domestic production (including renewable-energy sources) is limited by technical and economic constraints. This means acceptance of continuing risks of security of supply. Therefore the government sees it necessary to stimulate energy conservation; to increase or at least protect domestic energy production; to substitute capital or more secure forms of energy for oil; to reduce risks of energy imports where these are unavoidable; and to strengthen energy R&D expenditures.

German energy policy has developed a panoply of measures to reach these goals which will be described in detail. But overall energy policy in the FRG, despite (or because of) increasing problems and risks, has not changed its fundamental approach: to allow market forces to adjust to new situations as far as possible. This energy policy, relying on the market's adjustment mechanisms and intervening only where external effects would lead to intolerable results or where the adjustment process moves too slowly, seems to have been proven by the events of 1980 and the first half of 1981.

Energy Policy in the FRG before the First Oil Crisis (1973): Domination of Coal

It can with little exaggeration be said that for the last twenty years the FRG's energy policy was coal policy—especially policy to deal with the problems of the hard-coal-mining industry. Lignite (soft coal) production, which amounts to about 10 percent of the current PER, has, in contrast to hard coal, never created any substantial problems. This sector flourishes more than ever. There is no cheaper kind of energy. Reserves are tremendous and future problems seem to arise only from the question of how to distribute production among different purposes and consumers.

The hard-coal problem, however, dominated energy policy until 1973, and even since then most governmental actions and energy discussions have skirted coal questions. For more than twenty years energy policy in the FRG built one protective wall after another around the German coal industry. It is impossible to enumerate all the laws, measures, and regulations which were issued in the 1950s, 1960s, and 1970s to the advantage of indigenous coal. All of them had the objective of protecting indigenous hard-coal production or, later, at least securing a planned, phased retreat from it.

Historical energy-policy measures to protect indigenous hard coal include the following:

1. Measures which reduced direct competition from coal imports; such as a tax on coal from non-EC countries (1969), with a toll-free quota of about 5 to 6 mtce per annum (less than 10 percent of actual coal consumption).

2. Measures to render substitution by other kinds of energy more difficult, such as voluntary limits of imported-oil supplies in the early 1960s by the oil industry; taxes on light and heavy fuel oil (1960); and mandatory oil stocks to be held by refineries (ninety days of production) and importers of products (sixty-five days of imports). This last measure, instituted in 1965, was replaced in 1978 by a central organization holding inventories of the oil industry financed by a duty on every unit of oil product sold.

3. Measures to increase competitiveness of the coal-mining industry by grants, subsidies, and tax credits; examples of which would be: exemption from costs that were not caused by actual coal production (such as benefit payments for former miners); subsidies for shutdown of old coal pits and modernization of remaining ones; subsidies to transport coal or to produce electricity from coal; subsidies to equalize price differences for coking coal in the steel industry against the world-market prices; and subsidies for the construction and operation of facilities using coal, especially power stations and district heating units. Between 1960 and 1980 such subsidies amounted to over 30 billion DM.

4. Measures to limit or reduce the use of substitute fuels in the main economic sectors, and to restrict new power stations using fuel oil and natural gas since 1966.

In spite of all of these measures, the conditions for German coal could not be fundamentally improved. Competition became even more intensive because of the increasing penetration of the market by natural gas and the anticipated importance of nuclear energy. It was only in 1968 that the German government drew conclusions from the totally changed circumstances and passed a law (Kohleanpassungsgesetz) intended to reduce coal production to a competitive level. To reach this objective, political and financial assistance was given to concentrate the coal-mining industry into "economical units," which resulted, after 1968, in the foundation of the RAG (Ruhrkohle AG) with a 75 percent market share of coal production, the other 25 percent shared by five companies. Sales were concentrated in electrical generation and cokeries to assure security of supply in these main sectors and to maintain a secure core of energy supply.

In order to understand this focusing of energy policy on coal problems, it is necessary to keep in mind certain facts. First, huge coal resources in

Germany, as in many other industrialized countries, provided the energy basis for the industrialization process. As table 9-1 shows, in 1950 over 70 percent of the PER was supplied by hard coal (about 90 percent by all types of coal), and in 1960 hard coal was supplying 60 percent of the PER (75 percent for all types of coal). Second, after World War II, proven oil reserves seemed very small (at least to the public), the subsequent development of the world oil market seemed unpredictable, natural gas was nearly unknown in Western Europe, and hydroelectric power seemed already fully developed. Access to the German coal reserves to fuel economic reconstruction became one of the fundamental post-World War II objectives of energy policy of German and other European governments. Therefore, the nondiscriminatory access to German hard coal became in the early 1950s one of the vehicles for the European unification process.

As cheap hard coal was regarded as a fundamental basis for postwar reconstruction, the coal industry was not allowed to realize market prices. Instead, official prices were held on a level that neither reflected real scarcity nor allowed necessary investment and modernization. Consumers thus got the wrong signals and demand expanded. The coal-mining industry had to distribute its scarce supplies and asked for subsidies, which the government, feeling responsible for increasing production, granted. Consumers supplemented indigenous coal by imports on long-term contracts, which later rendered the adjustment process more difficult. Nobody realized (with the exception of a few oil companies) the fundamental changes, beginning in the middle of the 1950s, that had totally altered the energy-supply conditions of the Western world; replacing decades of scarcity and sellers' markets by decades of oversupply.

Huge crude-oil reserves (found in many cases before World War II) with negligible marginal costs were developed by the multinational oil corporations. A mandatory oil-import restriction program of the United States—the largest oil consumer at that time—pushed this oil to the European and Japanese markets at a time when German coal production was still increasing and energy imports were unnecessary. Low costs of production, economies of scale in transportation and refining, and intense competition by newcomers resulted in oil prices so low that European coal could not hope to compete.

But government policy cannot be described only from an energy-policy point of view. Surely, decades of controlled markets, scarcity, and rationing have deeply impressed the German political economy. Security of supply (normally linked with indigenous supply) has been and will remain one of the main energy-policy goals, but as always, other reasons and subjects have to be taken into account. One of the most important seems to be the fact that hard-coal production in Germany is concentrated in the Ruhr region, where, as politicians know, general elections are won in Germany. The well-being of the people working and living in this region has, therefore, always been of special interest in the political arena. Neither can one forget a very strong coal lobby, consisting of the mining industry, the mining-supply industry, the trade unions, and political parties. Especially after the second oil crisis, there are no real objections against domestic coal. The only difference between the political parties on the question of support of the coal industry is one of very minor degree.

Energy Policy in Other Energy Sectors to 1973

The description of energy policy as coal policy is obviously an oversimplification of the very complex energy policy in the FRG. Other main areas of political activity have been the publicly-controlled electricity sector,[2] and R&D, including nuclear energy. R&D activities have to be seen in the light of other political goals, such as the wish to bring German industry into a competitive position in a world market in which companies from other countries have benefited from military uses of nuclear energy and have been supported by their governments through huge R&D programs. Licensing procedures and enviornmental protection deserve mention as well. While these pose fundamental constraints on energy activities today, we need not treat them here in detail, because they have to be seen in a more general sense, not simply as energy-policy considerations.

Other historic energy-policy activities include subsidies for the conversion of coal-liquefaction plants to oil refineries (up to 1959); a 4-billion-DM subsidization of the search for oil and natural gas in the FRG (until 1964); subsidies and grants to German companies searching for oil outside the FRG; and the introduction of a general obligation to inform the government in advance of planned investments in production and transportation facilities within the energy sector. Most of these measures have been transitory, and have tried to compensate for structural disadvantages or to increase information for the government. In total, they seem to be of marginal importance in their effect on the total energy picture in the FRG.

Results of the Focus on Coal

While focusing on examples of government intervention in the coal industry, we must realize that the government has, by and large, refrained from intervention in the other energy sectors, though whatever is done in the coal sector inevitably affects the other sectors of the energy industry. For instance, the ambitious program to use hard coal in public utilities some years ago led to a less-intensive use of nuclear and lignite-fired power stations.

Government abstention from intervention in the noncoal energy sector may, therefore, more characterize German energy policy than the stubborn measures taken to protect domestic coal production—especially since this latter policy never prevented the downward adjustment process in the coal-mining industry, at best delaying it and smoothing the unavoidable friction.[3]

This is not just history. Many of the measures mentioned had originally been planned to operate only for a limited time, since both the coal-mining industry and many politicians had believed that the era of cheap oil, with its negative impact on the coal industry, would endure only a very short time. But the measures were continuously prolonged to deal with demands that rose to novel heights. And since, even with the profound price increases of the last few years, the ability of German coal to compete has continued to be a problem, security-of-supply considerations have remained paramount. Therefore, government intervention in the market to protect domestic coal has never ceased. So today many of the historical energy-policy measures described are still in force (though modified and partly reinforced). This is especially true (even with the revisions since January 1981) for the import restrictions on coal, the taxes on fuel oil, the direct subsidies to the mining industry, coal sold to the steel and electricity-generating industries, and for entry barriers to oil and gas for electricity generation.

1973 and Beyond

The events in 1973 totally changed the situation. Whether the dramatic changes were predictable or not is an academic question. The government of the FRG felt that the policies of protection of at least limited amounts of domestic coal production and the development of nuclear energy were justified. The downward-adjustment process for the hard-coal industry was immediately stopped and the necessary policy measures taken. With the comprehensive energy programs published in 1973, 1974, and 1977,[4] the German government officially followed the piecemeal approach, that is, energy policy by individual intervention, without explicitly taking into account the long-run aspects of energy supply and demand, the total system, or the interrelations among the different energy sectors.

Nevertheless, the German energy program is not a plan in the sense of a centrally planned economy, not even in the sense of an indicative (or *dirigiste*) plan of the French type. The German government has neither the instruments nor the intention to replace billions of private decisions by one central opinion, or even to heavily influence decisions in a desired direction. The German energy program only describes the government's interpretation of the energy situation, the expected developments in supply and demand (national and global), the problems that arise from these developments, and the policy measures that the government wants to take in order to prevent economic difficulties.

The Basis of Germany's Current Energy Policy

The first conclusion that the government drew from analyses of energy supply and demand in the FRG was that energy policy had to start from the structural facts and the prevailing conditions in our country.[5]

Since World War II the energy sector in the FRG had been characterized by both a constant increase of energy demand and a profound structural change brought on by interfuel substitution. With high economic-growth rates and increasing population, the PER rose by nearly 200 percent. But despite far-reaching coal protection measures, hard-coal consumption declined by one-third, to less than 85 mtce; the energy-market share of hard coal dropped from more than 70 percent to 22 percent of the PER. In the same period, oil consumption increased from 6 mtce to more than 200 mtce, its market share increasing from 5 percent to 55 percent. Natural gas, unimportant before 1960, by 1973 achieved an energy-market share of over 10 percent. Since domestic oil and gas reserves found and developed so far are rather small, the decisive result has been that energy supplies in the FRG have become increasingly dependent on energy imports (mainly oil). Net imports increased in 1973 to about 60 percent of the PER, whereas Germany in 1955 was a net energy-exporter—especially of coal—to its European neighbors.

This far-reaching integration of the German energy-supply system into the world market reflected consumer preferences for liquids and gases for heating and transportation (and the advantages of mechanical efficiency and versatility that these fuels possessed in supplying energy demand).

As an analysis of energy supply and demand and an evaluation of energy-policy options for the FRG shows, the fundamental characteristics of German energy supply cannot be changed in the short run without deep intervention in the system and drastic changes in consumer behavior.[6] Short-run price elasticities of energy supply and demand are rather low. The reasons are that all production and consumption processes need energy; that energy consumption is determined by long-lived appliances still in operation; that altering consumer behavior requires changing price expectations and/or preferences; that structural changes of the economy and in consumption patterns need time and are tied to huge investments to alter the capital stock; that alternative kinds of energy are available only in limited amounts and very often at increased prices; and that the development and penetration of new technologies to increase even traditional energy supplies very often show lead times of years and are constrained by environmental problems, high cost, and risks. Therefore, the German government still assumes that with growth rates of the Gross National Product (GNP) that are sufficient to reduce unemployment, finance our social system, and fulfill international obligations, energy consumption within the next decades will continue to increase even with a strong conservation policy. Though the rate of consumption is believed to be declining, and growth rates of the GNP are lower than previously assumed by the government (perhaps 2.5 percent per annum until 2000), the PER may reach 450-500 mtce in the year 2000.

Supply conditions also can only be changed step by step. The potential of domestic resources is limited and this creates further restraints for energy policy. The domestic-energy-supply situation can be summarized as follows:

1. Hydroelectric power has been used to its practical and economic limits.

2. German crude-oil reserves amount to only one-half year of consumption.

3. Natural-gas reserves developed to this point allow at most for sustaining present production (20 mtce) until the end of this century and no longer.

4. Domestic uranium reserves can supply only one unit of 1,200 MW for a lifetime of fifteen years.

5. Lignite reserves amount to 35 billion tons which equals present production (11 mtce) for three hundred years and would allow a large increase in production without resource restraints. But lignite is produced in open-pit mines in a heavily populated area of Western Germany and it has to be assumed that a limited increase in production (10 to 20 percent) would describe the highest level of environmentally acceptable production. This would only meet a small percentage of the additional demand.

6. Hard-coal resources are very large. They add up to some 230 billion tons. But with current technology, recoverable resources are estimated at only 24 billion tons of coal equivalent (btce). Economically producible reserves, though not definite since they are a function of costs and prices, are estimated to range between 3 and 9 btce. Therefore, with respect to the resource base alone, it is doubtful whether it would have been better to protect the domestic coal-mining industry more than we did. I suppose that no government would have been able to pay the bill to maintain, for example, a 150-mtce coal-production capacity in Germany. In this regard we should remember that even after the shutdown of one hundred coal pits, a two-thirds reduction in manpower, and far-reaching modernization and concentration; even after crude-oil prices in Germany have increased by more than 800 percent; German coal still needs heavy subsidies—this year (1980) alone, more than 6 billion DM. There are doubts as to whether even a slightly in-

creased coal production over today's level will be possible. The resource base is tremendous, but reserves amount only to a small fraction of resources, and even a subsidized mining industry in the FRG faces more and more problems with respect to skilled labor and environmental considerations. Indeed, the maintenance of actual production (90 mtce) will require substantial support from the government.

Even under relatively optimistic assumptions, renewable types of energy will play only a limited role in the German energy supply until the end of this century. Under natural conditions, and at their present technical as well as economic stage of development, it is believed that renewable-energy sources could supply at most 10 percent of demand—with much of this coming from hydroelectric power and heat-exchange pumps.

This all leads to the conclusion that energy needs in our country for the next decades will increasingly have to be supplied by foreign countries. This fact determines the direction and type of energy-policy measures. It means that only a limited amount of supply security can be reached by measures to sustain indigenous production. Energy policy will have to consist of measures to increase the energy efficiency of our present systems and to reduce the risks of remaining energy imports.

The government of the FRG was not concerned about our growing dependence on energy imports as long as the world energy market was characterized by oversupply, as long as new energy provinces continued to be developed, and new kinds of energy (such as natural gas) began to penetrate or were expected (such as nuclear power) to enter the market. Above all, there was little worry as long as the extraordinarily flexible system run by the giant multinational oil companies existed, since this seemed able to overcome even severe supply interruptions. Since 1973, we have obviously been faced with a totally different situation. The era of cheap, abundant oil has been consigned to history. More of the power to dispose of the resource base has been transferred to the producing countries, who use their resources in their own economic and political interests. These interests often do not coincide with our own.

The framework of energy supply in the FRG, as in all oil-consuming nations, has totally changed. No one can be sure that supply conditions will not worsen in the future; this is another basic assumption of recent energy policy in Germany.

The General Philosophy of Energy Policy in the FRG

One could believe that the dramatic changes in the world energy markets, the increasing political influences that substitute for economic considerations, the high and continuing dependence of our energy supply on frictionless world markets, the increasing threat of interruption of supply, the danger that further price shocks may destroy our economies and endanger our welfare—one could believe that all of these factors would recommend a total change in our existing energy policy system. In addition, we are facing large and increasing uncertainties in nearly all aspects relevant to the development of the energy sector; uncertainties regarding: the further development of energy consumption with respect to changing macroeconomic conditions such as growth rate and structural changes within industry resulting from the impacts of increasing prices and changes in behavior and preference; the supply curves of the various kinds of energy; and technical progress, in which environmental constraints and political influences become paramount.

Now, more than ever, the German government (in contrast with a minority faction in the left wing of the Social Democratic Party) adheres to the philosophy that an optimal solution of our energy-supply problems can only be expected within the framework of a system of decentralized decisions, a framework compatible with the social-market economy. Energy policy in the FRG plays no isolated role. It is part of economic policy and derives its goals from higher-ranking policy objectives such as economic growth. This requires a maximum of dynamics and adaptability in the energy sector. Governmental interventions are often considered less flexible and reversible than decisions resulting from market forces.

The competitive pressure of the market can force corrective adaptation in the case of faulty decision making (the current rearrangement of refineries serves as a good example). Governmental authorities can, however, only hush up and postpone the consequences of false decisions through subsequent intervention. Adaptability cannot be administered. It can only be realized through a market system based on competition. The short-run rationality of the market is often raised as an argument to intervene in the energy sector. It is argued that the market is incapable of considering long-run price trends accompanying current investment decisions. That argument is wrong. The decision horizon of the politicians with respect to the next election might be shorter than that which firms apply to their long-run decisions. With regard to these insights, therefore, it is the objective of the German energy program to strengthen market forces and to intervene only when market imperfections lead to politically unacceptable results or where market forces need support or correction because of high risks, long lead times, or external effects.

A market-oriented energy policy is not a policy of laissez faire. It does not lack a conscious policy design. It supports market forces, preserves as far as possible the sovereignty of consumers, and leads, in principle, to an acceptance of all risks by private decision makers. The government intervenes primarily by means of setting an adequate data framework and ensuring that prices are the central steering mechanism, prices that reflect real scarcity and allocate goods and services in an optimal manner.

These considerations resulted in the German energy programs. The general objective is to supply energy needs at the lowest long-run social costs with a market system and with as few governmental interventions as possible, but implicitly taking into account the structural problems and necessities that face energy supply and demand in our country. Such an academic, abstract formula had to be translated into practical energy-policy goals. The goals described in the program are that energy policy should secure: a sufficient energy supply for all consumers in all areas of Germany at reasonable prices; medium-term and long-term security of supply; a reduction of the social costs of energy supply; and the internalization of environmental-protection costs and the costs of interruption of supply.

Strategy Issues and Policy Instruments

These objectives were transformed into main strategy issues and challenges as follows:

1. Reduction of the oil share of the PER.
2. Enhanced diversification of supply.
3. Stabilization of domestic energy production.
4. Support of nonnuclear R&D.
5. Development of a workable crisis-management scheme.

There are five groups of measures to implement these policies:

1. An extensive program to save energy (conservation).
2. Promotion of domestic energy resources.
3. Increasing nuclear energy.
4. Development and marketing of substitutes for oil.
5. Crisis management for cases of interruption of supply.

Conservation measures included:

A retrofit tax-credit and grant program for homeowners under the "Housing Modernization Act. This program (including subsidies for new technologies and district heating) began in July 1978 and will run five years at a cost of 4.35 billion DM. A follow-up is in preparation.

A requirement that heating systems have to be serviced regularly. In the case of new installations, devices to regulate heat supply are mandatory.

A voluntary commitment by appliance manufacturers for labeling how much energy is used.

A voluntary program by automobile manufacturers to improve fuel economy by at least 10 percent by 1985.

An amendment to the Investment Allowance Act of 1975 to provide a 7.5 percent allowance for investment in combined-heat-power (cogeneration) plants, industrial-waste heat systems, and district heating systems, plus grants of 700 million DM for combined heat cycles.

Confirmation of government intent to continue promotion of district heating and construction of coal-fired power stations after the "Investments for the Future" program (600 million DM) expires in 1980.

An advisory service for private consumers and small and medium-size industries, plus advertising and information programs.

Grants to accelerate marketing of new energy-saving technologies.

A proposed amendment to the "Law on Energy Conservation" that allows consumption-oriented invoicing of heating costs and the possibility of limited requirements for thermal insulation and heating plants in existing buildings.

New mandatory standards raising the level of thermal-insulation and heating-system standards for new houses. The possibility of retrofitting existing houses and introducing standards in these buildings (which make up the largest share of potential conservation for the future) is being discussed at the time of this writing.

An increase in investment in R&D funding of new conservation technologies.

Measures to promote domestic resources and reduce oil's share of the PER include:

Grants and subsidies to the German coal-mining industry.

Subsidies to public utilities for the use of 33 million tons of domestic hard coal (average) per year, and to build up new coal-fired power plants. These subsidies amounted to nearly 9 billion DM between 1974 and 1977. Since 1977, electrical consumers have financed these subsidies by regionally differentiated duties on their electricity bills (about 5 percent average). This amounts to about 3 billion DM per year.

Strong restrictions on the additional use of oil and gas in electricity generation and subsidization of the substitution of coal for oil and gas in the electricity sector.

Sustaining the subsidies to coking-coal users in the European steel industry, which reduces the costs of German cooking coal for these consumers to world-market prices. These subsidies amounted in 1980 to about 2.0 billion DM.

R&D programs for the development of new coal-mining technologies as well as new coal-using facilities. The costs for this between 1977 and 1980 have amounted to 900 million DM.

Steps taken to increase nuclear energy include:

Strong support of an originally ambitious, but now much reduced, nuclear-development program.

Development of plans for a central storage and reprocessing facility at one site (Gorleben). After significant opposition to this facility and site, the government, though being convinced that the original project is secure and feasible, has followed a mixed strategy allowing expanded onsite interim storage; several central interim-storage facilities; and smaller reprocessing plants (not in Gorleben). But the government still supports the idea of a single waste-disposal center.[7] This new concept was agreed upon in the spring of 1980 between the Chancellor and the prime ministers of the Länder, and shall give policy direction to the courts—which, because of the unsolved problems in this regard, have heavily delayed commissioning of new nuclear plants. This has led to a de facto moratorium on nuclear construction in Germany.

Subsidies for the development of advanced nuclear technologies (breeder and high-temperature reactors), but at a reduced level.

A reduction in oil's share of the PER will be achieved by:

Not only the subsidies mentioned for domestic coal, the increased use of district heating, and conservation measures (including heat pumps), but the limited opening of the coal-import market by an immediate increase in the toll-free contingent of industry (until 1983, if oil and gas are replaced).

Announcement of a large-scale synthetic-fuels and gases program (January 1980), which shall demonstrate the feasibility of the projects and improve economies of coal liquefaction and coal gasification on a very large scale, and which will be subsidized (30-50 percent) of investment) by the government.

Doubling the tax on heating oil.

Strong control of diversification of gas and oil sources by grants to the German oil-exploration group, DEMINEX.

An energy-related R&D program (with subsidies of 6.5 billion DM since 1977).

A crisis-management program has been developed for interruption-of-supply-situations (in addition to still-existing measures like mandatory oil stocks to be held by refineries and product importers). This program includes:

Building up government-owned crude-oil stocks to 8 million tons.

Building up a government-owned hard-coal stock of 10 mtce.

Laying the groundwork for an energy-security law for crisis-management in cases of interruption of supply that are not governed through the IEA system.

Evaluation of the Energy Program

This is a fairly ambitious program. Though coal-related measures may still outweigh conservation, German energy policy has become much broader. Despite all of its transgressions, German energy policy still follows the general philosophy of letting market forces work as long and as far as possible. This means that government intervention in private decisions shall remain the exception. One remarkable result of this philosophy—especially in relation to other European countries—is that the government in the FRG has never tried to enforce its energy-policy goals through nationalization or through the setting up and supporting of government-owned companies. The fact that the government owns the largest German energy company, Veba, by 43 percent is a historical relic, and government has given up its majority in that case by selling part of the shares to the private sector. It has never been used for energy-policy purposes. Veba is managed like a private company—not necessarily in line with public interests.

The same is true for the numerous publicly owned companies in the electricity and gas sectors, or public ownership of coal-producing companies like Saarbergwerke. Market entry in principle is free, and supply and demand are steered by prices that reflect, with few exceptions, the conditions of world markets. Energy policy is not seen as a means of income distribution or as an instrument of social policy. Social problems of low-income groups (those which suffered from the tremendous oil-price increases in 1979) were not remedied by a price-fixing policy, but have been corrected by direct payments to this group of our society. The most important expression of this general energy-policy philosophy is that price increases, even if they create huge problems for consumers and industries, have not been prevented by the government. They are officially seen as reflections of real scarcity and are expected to reduce demand and to increase supply. The government only tries to prevent monopolistic profit by means of a sharp anti-trust policy.

Yet we would confess that even the German energy-policy system is far from the liberal ideal. Reality is a mix of market system and government control that could, without a doubt, be improved. For example, the far-reaching coal-protection measures, with their impact on other sectors, have (correctly) been criticized in the past. But on the other hand, we have to see that the internalization of external effects can only be achieved under a political aegis. It is also necessary that government intervention be legitimate. Often domination by special interests, interference by nonenergy policy issues, or bureaucratic arrogance underlie energy-policy measures or proposals.

How shall we evaluate our energy-policy system in the light of the actual development of the German energy market as described? Did our energy policy totally miss its objectives? Did we go in the right direction? Did we take the adequate measures with sufficient urgency and intensity? Surely it is too early to judge with any certainty.

In any case, I do not believe that the development of energy supply and demand in the year 1979 adequately reflected the adjustment process that is still taking place. A detailed analysis shows that energy consumption in our country in 1979 was influenced by a number of unique facts that did not allow for simple extrapolation into the future. Consumption was strongly influenced by a comparably high (for Germany) economic-growth rate (4.4 percent) with high growth rates in energy and oil-intensive branches like the chemical industry; by an unusually cold year; by higher stockpiles of fuel; and by larger numbers of new cars and centrally heated housing units.

The increase in oil consumption in 1979 of 1.3 percent was largely determined by the growth of the chemical industry (10 percent), the trucking industry, and to a small extent by an increase in the number of private cars. While the number of cars increased by 4.6 percent, gasoline consumption was only 1.3 percent higher than in 1978. All other oil-consuming sectors remained constant or decreased.[8] Since 1973, the oil share of the PER has dropped from more than 55 percent to 51 percent (see table 9-1), while our economy has grown by 15 percent. The number of centrally heated housing units increased by 40 percent and the number of new cars by 33 percent. In 1980, the PER was reduced by 3.5 percent, and oil consumption by 11 percent—leading to an oil share in the PER of 47 percent. The same reduction of the PER and oil consumption has taken place in the first six months of 1981. We can expect that the PER in 1981 will reach 375 mtce (lower than 1973) and that oil consumption will decrease to the level of the late 1960s.

This shows that consumers react remarkably well and that the adjustment process proceeds. This adjustment process only began with the new oil-price increase in 1979, which, reinforced by the factual devaluation of the deutsche mark against the dollar, increased the real energy prices for the German consumer considerably and changed his price expectations fundamentally. As the adjustment process is to a large extent linked to new investment or reinvestment in production facilities and appliances as well as to new technologies, an acceleration of these effects over time can be expected.

In our government energy projection in 1977, we therefore assumed that the ratio between economic growth and increase in the PER would be reduced from about 1.0 (in the period 1960-1973), to 0.8 by 1985, and to less than 0.5 by the year 2000. The share of oil consumption in this projection was estimated to decrease to less than 30 percent. As the price assumptions underlying this estimate (doubling by the year 2000) have been superseded, and as new conservation policies have been implemented, the results may improve even higher. According to our recent estimate,[9] energy consumption in the household, commercial, and traffic sectors has peaked, and the PER will only increase at a low rate because of the growth in industry and the necessary losses to produce electricity, and will reach a plateau in the 1990s.

Remaining Energy Problems

There remain many problems. Among these is the problem of windfall profits for indigenous producers (and their competitors), the problem of how to finish the debate about nuclear energy, and the problem of the degree to which the domestic coal industry should be protected.

Germany too has a problem with windfall profits, in which domestic producers profit (in 1981 about 2 billion DM) from increasing world-market prices without any effort of their own. A taxation of windfall profits raises allocative and distributive questions. Among these are the questions of: What kinds of energy production shall be taxed to what degree and for how long? What about windfall losses? Will the have-nots be withdrawn from the market? What are the results of a concentration process with respect to competitive prices? In Germany the idea of taxation has been given up—at least for the moment. Royalty payments have increased from 5 percent until 1976 to 22 percent, and the new mining law will give (from 1982 on) the possibility of an increase in royalties to 40 percent. Unfortunately, the actual rules might lead to heavy problems for marginal producers as long as basis for the taxation is the market price and not the net profit of production.

Nuclear energy once promised to become one of the main options to reduce our oil dependence. With subsidies that have amounted to about 20 billion DM since 1956, and other public supports, nuclear energy should have developed significantly. The energy program of 1973 expected to have 45-50 GW of installed nuclear capacity by 1985. But public acceptance (or nonacceptance) have together with courts and lack of political direction increasingly delayed the construction of nuclear power plants, so that the nuclear program has been (and continues to be) reduced. By 1985 we can expect at best to have installed a nuclear capacity of about 17 GW. The nuclear debate transcends all political parties. Part of our society rejects nuclear energy totally. Another part concedes a role for this kind of energy if all conservation and coal-use possibilities have been tried.[10] Since the coal question seems to be moving toward a solution (to be discussed), the government is again searching for ways to overcome the nuclear-acceptance problems, which are seen to stem mainly from problems with reprocessing and the final storage of nuclear waste. More and more it is believed in the FRG that the time for further discussion is past and that only political decisions can resolve the situation. But whether public acceptance can be forced must remain unknown. A majority of our society seems to agree that at least a limited use of nuclear energy is necessary in our country and that the risks are tolerable.[11]

A remarkable step in the direction of solving the coal industry's problems has been the opening of the market for coal imports (from January 1, 1981) by guaranteeing certain markets for the German coal-mining industry. Public utilities and the electricity-generating industry have more-or-less voluntarily agreed to increase their consumption of domestic coal gradually to 45-50 mtce by 1995—by about 50 percent. The government will subsidize (if necessary) part of this consumption through funds financed by electrical consumers, and take as well the necessary measures to at least sustain coal production at today's level. This gives the coal mines the guarantee that more than 50 percent of the expected production until 1995 can be sold at cost to utilities. Another 35-40 percent is guaranteed by subsidized deliveries to the steel industry until 1988. Public utilities and industrial-electricity producers will, at the same moment, get the right to import coal in a firm, gradually increasing amount relating to the additional consumption of German coal. From 1981 to 1995, this may total as much as 120 mtce. Industrial consumers will be allowed to import coal in graduated amounts (20 mtce annually by 1985, 40 mtce by 1990, and 60 mtce by 1995) for heating purposes. In addition, the steel industry will be allowed to import 3 mtce annually. Imports of 5 mtce for the production of synthetic fuels are foreseen as well. Lastly, it will be possible to increase these quotas by 50 percent if necessary.[12] Thus, one of the most dirigiste and protectionist sectors of the German energy sector will be shut down after a quarter of a century.

A market-oriented energy policy is a continuing challenge. Our government until now has clearly followed this path. But temptation and external pressures to become more dirigiste, to reduce private responsibility and substitute a bureaucracy for market mechanisms, are very great. Finally, the integration of our energy policy into international and supranational agreements and arrangements raises continuing internal problems for our system. Perhaps this is one of the most important reasons for the evident, though not official, aversion of our main political leaders for a European energy policy. To go more European in this sense, under the existing (and probably future) conditions, would necessarily result in chaging our energy-policy system—against the wishes of the preponderant majority in the FRG—in the wrong direction. We hope it is possible to resist what we believe is a wrong direction, or misstep, on the road to European integration.

Notes

1. An updated edition of the FRG energy program is planned for the end of 1981.

2. For detailed information see: W. Schulz, *Ordnungsprobleme der Elektrizitätswirtschaft. Eine Überprüfung wettbewerbspolitischer Argumente mit Hilfe eines Simulationsmodells* (Munich, 1979).

3. German hard-coal production decreased from a 1957 peak of more than 155 mtce to less than 100 mtce. Employees were reduced from more than 600,000 to less than 200,000. In 1957, about 150 pits produced, on the average, 1 mtce a year. In 1978, only 40 pits produced more than 2 mtce, with the highest rate of productivity in Western Europe. Nevertheless, production costs to 1980 have been twice as high as the price of imported coal.

4. Ministry of Economic Affairs, Bonn, "Energieprogramm der Bundesregierung," 26. December 1973; also 1st rev. ed., 1974; 2d. rev. ed., 1977.

5. Conducted by the Institute of Energy Economics at the University of Cologne and the Energy departments of two other German research institutes.

6. DIW/EWI/RWI, *Die zukünftige Entwicklung der Energienachfrage in der Bundesrepublik Deutschland und deren Deckung. Perspektiven bis zum Jahre 2000* (Essen, 1978).

7. In addition, storage of spent fuel elements shall be investigated.

8. In the FRG, only 8 percent of electricity is oil-generated as opposed to 56 percent in Italy and 53 percent in Japan.

9. Ministry of Economics, "Dritte Fortschreibung des Energieprogramms der Bundesregierung" (Bonn: November, 1981).

10. This is the main result of a so-called Kernenergie Enquête-Kommission on Future Nuclear Energy Policy of the Parliament, which should answer the question of whether Germany should go nuclear or not. See Bundestagsdrucksache 8/4341 vom 27.6.1980.

11. This is the position of the new government, which can rely on the official declarations of all the major parties, the industrial federations, and the trade unions.

12. "Zweites Gesetz zur Änderung energierechtlicher Vorschriften vom 25 August 1980," Bundesgesetzblatt I, p. 1605ff.

References

Krüger, M., ed. *Energiepolitik, Kontroversen, Perspektiven*. Cologne: 1977.

Meyer-Renschhausen, Martin. *Energiepolitik in der BRD von 1950 bis Heute*. Cologne: 1977.

Michaelis, Hans. "Europapolitik und Energiepolitik," in *Zeitschrift für Energiewirtschaft*, Heft 1, 1979.

Mönig, W., Schmitt, D., Schneider, H.K., Schürmann, H.J. *Konzentration und Wettbewerb auf dem Deutschen Energiemarkt*. Munich: 1977.

Rühle, H. and Miegel, M., editors. *Energiepolitik in der Marktwirtschaft*. Stuttgart: 1980.

Schmitt, D., Schneider, Hans K., Schürmann, H.J. "Nach Harrisburg und Teheran: Eine energiepolitische Bestandsaufnahme," in *Zeitschrift für Energiewirtschaft*, Heft 2, 1979.

Schmitt, D., Schürmann, H.J., "Grundlagen einer zukunfts-orientierten Energiepolitik," in P. Harbusch and P. Wick, editors. *Marktwirtschaft*. Stuttgart: 1975.

10 Italian Energy Policy

Umberto Colombo

As a result of the second oil crisis, many OECD countries experienced decreasing or negative rates of growth as well as increasing balance-of-trade deficits. An increase in the price of oil (a 130 percent rise over eighteen months for Arabian Light) rather than a physical lack of supply was largely responsible for the second oil crisis. In Italy, economic growth remained strong during the crisis; however, Italy's balance of trade deteriorated rapidly.

Clearly, oil is the dominant source of world-energy supply: it supplies 45 percent of current world-energy demand and about 51 percent of the energy demand of the seven main industrialized countries of the OECD (the United States, Canada, Japan, France, West Germany, Great Britain, and Italy). All of the OECD industrialized countries including those with notable oil reserves depend on oil supplies from abroad. Among the industrialized countries not possessing their own oil reserves, Japan has the highest import rate, followed by West Germany and Italy. In 1979, Japan imported 99.8 percent of its total oil requirement, West Germany imported 98.9 percent, and Italy imported 98.7 percent (see table 10-1).

A closer consideration of the energy situation in Italy indicates that its dependence on hydrocarbon fuels in 1979 amounted to 84.1 percent of total energy consumption. In addition, dependence on the Organization of Petroleum Exporting Countries (OPEC) reached 80 percent of its total oil imports during the same period. Italy still contributes significantly to its own supply of natural gas and obtains a large part of its imported natural gas from sources other than OPEC suppliers. Finally, Italy's per-capita energy consumption (2.60 tons of oil equivalent in 1979) is substantially lower than the overall OECD average (5.02 tons of oil equivalent per capita in 1979).

The Evolution of Energy Consumption in Italy

In 1955, at the end of the postwar reconstruction phase, hydrocarbon fuels accounted for 43.8 percent of Italy's energy consumption. Hydroelectric sources were also a major factor, accounting for 31 percent of supply, with the added advantage of restricting both in volume and in currency the country's dependence on imports. During this period, far-reaching changes in Italy's productive structure began to take place. Evidence of these changes is seen in the increase in the domestic industrial contribution to the GNP,

Table 10-1
Energy Consumption by Major Countries and World Groupings, 1979
(millions of tons of oil equivalent)

Type of Energy	*Japan*	*United States*	*Soviet Union*	*Italy*	*EC*	*OECD*	*World*
Solid fuels	58.6	384.1	342.5	11.7	214.7	757.3	1,979.0
	(15.4%)	(20.2%)	(29.8%)	(7.9%)	(21.3%)	(13.4%)	(28.4%)
Oil	265.4	862.9	441.0	101.5	550.8	1,981.2	3,120.0
	(69.7%)	(45.5%)	(38.4%)	(68.5%)	(54.7%)	(50.7%)	(44.8%)
Dependence on imported oil	(99.7%)	(44.2%)	—	(98.7%)	(87.0%)	(67.0%)	—
Natural gas	22.1	498.8	307.0	23.0	176.8	764.0	1,297.0
	(5.8%)	(26.3%)	(26.8%)	(15.6%)	(17.6%)	(19.5%)	(18.7%)
Hydroelectric-geothermal	19.9	80.1	45.0	11.3	31.4	271.9	411.0
	(5.2%)	(4.2%)	(3.9%)	(7.6%)	(3.1%)	(6.9%)	(5.8%)
Nuclear	14.7	72.2	12.5	0.6	32.7	136.7	155.0
	(3.9%)	(3.8%)	(1.1%)	(0.4%)	(3.3%)	(3.5%)	(2.2%)
Totals	380.7	1,898.1	1,148.0	148.1	1,006.4	3,911.1	6,962.0
	(100.0%)	(100.0%)	(100.0%)	(100.0%)	(100.0%)	(100.0%)	(100.0%)
Tons of oil equivalent per capita	3.29	8.60	4.80	2.60	3.86	5.08	1.69
Percentage of world consumption	5.47	27.27	16.49	2.13	14.46	56.19	

rising from 32.9 percent in 1960 to 37.3 percent in 1973. A corresponding decrease is found in the domestic agricultural contribution to the GNP, falling from 13.6 percent to 9.3 percent during these same years.

With this broadening of the industrial base, a marked increase in Italy's GNP as well as a marked increase in energy consumption developed. In real terms, the GNP increased by approximately 5.2 percent per year from 1960 to 1973, and energy consumption increased by 8.1 percent per year. These increases were greater than in many other industrialized countries. One major reason for the greater increase in energy consumption in Italy as compared with the increase in income relates to the importance that basic industries acquired during these years. Such industries are, of course, energy-intensive. The exceptional expansion of the transport sector, related to the increase in both national income and industrial output, is another factor that contributed to a high level of energy consumption.

Parallel to its ever-greater consumption of energy, Italy increasingly resorted to petroleum. By 1973 petroleum supplied 75.3 percent of its energy needs. With the rise in the price of oil and other raw materials, Italy experienced a slowdown in the growth of its economy from the previous yearly average of about 5.2 percent to about 2.2 percent for the 1974-1978 period. In addition, the inflation rate increased and the balance-of-payments deficit worsened.

Contemporary with these changes, a modification in the pattern of energy use came about in regard to primary-energy sources. With a decline in oil consumption from 102 million tons of oil equivalent (mtoe) in 1974 to 99 mtoe in 1978, and a stationary demand for solid fuels and for electric energy from primary sources, natural-gas consumption increased from 16 mtoe to 22 mtoe concomitantly with the entry into service of the great gas pipelines of the Soviet Union and the Netherlands.

Another modification in energy use, though to a smaller degree, was brought about by numerous forms of energy saving, especially in the industrial sector. If one analyzes the patterns of total-energy and electric-energy consumption per unit of value added to industrial products at constant prices for 1970, it is evident that efforts to conserve energy following the 1973 oil crisis made significant reductions in total specific forms of consumption. However, electricity consumption per unit of value added to industrial products has not varied greatly, a circumstance that points to the higher level of optimization that had already been achieved in the use of electricity.

Impact of the Second Energy Crisis in Italy

In 1979, after two years of only moderate growth, the Italian economy expanded rapidly. Italy's GNP rose by 5 percent in real terms, against an average OECD growth rate of 3.4 percent. The economic expansion derived largely from an increase in all factors of demand: an increase of 4.7 percent in final domestic consumption; an increase of 4.7 percent in investments; and an increase of 8.7 percent in exports. Thus in Italy, where oil accounts for roughly 69 percent of energy consumption, economic growth was great, despite the rapid rise in oil prices.

At the same time, nevertheless, there was a steady deterioration in the balance of trade. The sharp rise in exports was more than counterbalanced by a growth in imports. After only a minor deficit in 1978 of 363 billion lire, 1979 ended with a deficit of 4,726 billion lire (roughly 5 billion U.S. dollars). A major cause of this deterioration was the cost of oil. The oil deficit went from approximately 8,000 billion lire in 1978 to 10,627 billion lire in 1979, resulting almost exclusively from increases in the purchase price of oil (a 31.7 percent average increase in 1979). For the first nine months of 1980, the trade deficit came to 13,612 billion lire, approximately ten times the deficit for the same period in 1979. This gloomy situation was the result of a negative balance of trade in the energy sector of 12,891 billion lire, of which 11,897 billion lire were attributable to oil imports. Additional factors concern economic difficulties encountered by other sectors of the Italian economy.

One of the major factors is the inflation rate. In 1979 retail prices in Italy rose by 14.9 percent compared with an OECD average of 10.9 percent. Apart from the steep rise in oil prices—a factor common to all industrialized countries—this inflation rate was the result of endogenous impulses typical of the Italian economic system: shortages of raw materials and rigidities in productive capacity. The inflationary phenomenon continued with particular intensity during the early months of 1980. In July, for example, the inflation rate increased by 20.6 percent over the rate for the same month in 1979.

The worsening of the energy and economic situations prompted the Italian government to step in with the measures it believed were appropriate. However, such measures were not often implemented; and when they were, they failed to provide the hoped-for results. In the energy field, policy measures turned out to be little more than sporadic gambits. An energy-saving plan, submitted to Parliament by the Minister of Industry in June 1979, has become a dead letter. Nor has the nuclear-power-plant program begun.

In September 1979 a Decree Law was proposed for the purpose of controlling the overall period and number of hours per day in which heating could occur. This measure was recalled because of opposition to another point in the same bill—the establishment of an energy-emergency fund that would, among other things, finance energy saving. Despite later attempts to reintroduce the measure, it has not obtained parliamentary approval.

Another important step in energy policy was the establishment in December 1979 of the Permanent Technical Committee for Energy (CTPE), chaired by the Minister of Industry and composed of the chairmen of the national power boards—CNEN (nuclear), ENEL (electrical), and ENI (hydrocarbon)—and a panel of experts, three on energy problems and one on legislative problems. The aim of the committee is to coordinate the work of the national boards, to put forward technical and administrative suggestions with regard to energy, and to ensure that decisions are implemented.

In short, since 1979 there have been many attempts to implement an energy policy. While some decisions have been made, many have been postponed or defeated by various political vetoes.

In the field of economic policy, action was taken toward the end of 1979 to bolster domestic demand by means of an expansive budget policy in order to counterbalance the slump in foreign demand and to permit the investment cycle to develop fully. This policy far exceeded even the highest expectations. It gave rise to a large increase in industrial production, domestic demand, investment, and employment. Thus, as noted, the trade deficit grew. Aside from a consequent increase in the demand for costly imported oil, domestic investment and demand for consumer goods grew in excess of domestic production, which was not sufficiently elastic to meet such an expansion.

The high level of domestic consumption added to growing inflation, paving the way for increasing losses in competitiveness in foreign markets. Therefore, the government stepped in once more with restrictive measures, aimed this time at attaining two objectives: recovery of competitiveness in foreign markets in order to eliminate the deficit in the balance of trade, and containment of domestic demand in order to reduce the inflation rate.

Institutional Aspects of Italian Energy Policy

In Italy the National Energy Plan, which is periodically updated, provides the most authoritative guide to energy policy.[1] Current procedure for drawing up and obtaining approval of the Energy Plan constitutes a series of technical and political stages.

Initially the Energy Plan is developed by the Ministry of Industry and Trade. For this purpose the ministry calls in the major public and private bodies concerned with the energy sector for consultation. The proposal is then examined by the Permanent Technical Committee for Energy (CTPE). Next the proposal is examined and approved (or disapproved) by the Council of Ministers; if approved, the proposal is debated in Parliament. Finally, the Energy Plan is examined by the Inter-Ministerial Committee for Economic Planning (CIPE), which considers Parliament's intentions and concerns for the Plan and issues an implementation resolution for the Plan. The CIPE is composed of all ministers concerned with economic matters or problems connected with the energy and planning sectors.

All measures outlined in the Energy Plan and approved by the CIPE that involve the creation of laws (institutional changes, or financing over a period of time) require additional approval. Procedurally, the minister under whose jurisdiction the bill falls draws up the legislation. Other interested ministers become involved, then the Council of Ministers, and finally Parliament.

The main shortcomings found in Italy with regard to the elaboration of an energy policy can be summed up in two points. First, there is a lack of continuity in political measures because of frequent government crises and erratic parliamentary consideration of energy problems. Second, there is the difficulty encountered in implementing the energy policies approved in general terms by Parliament and decided on by the CIPE. This stems from the absence of a single, coordinated direction at both the executive and political levels.

Some observers hope to ameliorate these faults by setting up special bodies. On the political level such a body could be a committee established within the framework of the CIPE. It could be chaired by the Prime Minister and would consist of all the ministers concerned with energy. The task of the body would be to define overall energy policies, and to establish lines along which coordinated action could be taken by all the departments of public administration. On the executive level, Italy should consider the creation of a Ministry of Energy or an Office of Energy Problems headed by a commissioner. An alternative solution might consist of concentrating all the energy tasks currently performed by a range of ministers under the Minister of Industry and Trade.

Italian Attitudes toward the EEC and the IEA

The limits encountered with regard to energy-resource availability and the objective of maintaining its role as an industrialized nation have induced Italy to support all international efforts aimed at establishing an energy policy common to energy-consuming countries. These efforts are aimed in particular at the following: boosting the energy-consuming countries' bargaining capacity in relation to the energy-producing nations; creating a combined approach to ventures involving political difficulties; calling for commitments in policy areas too complex for any one country to handle; and harmonizing the national energy policies, thus facilitating coordination of overall objectives of economic policy. Unfortunately, the energy-policy goals set over the last years by the European Economic Community (EEC), resulting from a compromise among differing national needs, do not satisfactorily reflect Italian aspirations.

One of the cornerstones of EEC energy policy is the reduction of the ratio between the growth rate in consumption of primary energy and the increase in Gross Domestic Product (GDP) by energy-saving measures. The 0.7 average value established by the EEC for this ratio by 1990 may seem an ambitious one for a country like Italy, whose energy/GNP elasticity ratio has been constantly above 1.0 for the last three decades. However, recent data indicate that this target could well be reached with conservation policies, based on the experience of other European countries.

Other EEC energy-policy goals are the reduction of the share of oil as part of the member countries' energy budgets, a share that should fall from the present 55 percent to 40 percent by 1990; the increased use of coal and nuclear energy to produce electricity; and the growing use of renewable energy sources. These goals are considered of special importance by Italy, which trusts that the EEC will use all the means at its disposal—financial aid, price policy, tax inducements, and technological development—to promote the coordinated attainment of these objectives.

In the field of renewable energy sources, it is hoped that the EEC can take measures to start pilot projects, resolving the current proliferation of free ventures that marks the sector today and encouraging the concentration of efforts on solutions that are both more qualified and more effective from the technical and economic standpoints. In this connection, new life should be given to the Common Research Centers, set up mainly to develop nuclear applications, concentrating efforts on the safety problems of nuclear energy and gradually being converted into research centers for renewable energy sources.

The International Energy Agency (IEA) is the second international body of great importance to Italy. Founded in 1974, the IEA has fully achieved several of its minor objectives such as the acquisition of a wide-ranging information system on the international oil market, the establishment of a broad cooperative effort in the energy-saving sector, and the development of alternative energy sources. In this last field, however, the concern manifested by Italy when the IEA was founded as to the limitations of its institutional mandate regarding nuclear energy has proved justified by the course of events.

In substance, the IEA has restricted the range of measures likely to produce meaningful short-term alternatives to oil. In the case of the nuclear sector, the IEA has limited itself to retrospective reviews and projections rather than going into studies that could provide direction and impetus. In particular, areas of enormous importance such as exploring for uranium, perfecting emergency mechanisms to set up uranium stocks, disposing of radioactive waste—areas in which Italy invited the IEA to take an interest—have been almost totally ignored. Moreover, it is as yet impossible to assess the work of the IEA on its most pressing institutional duty—the attempt to assemble a solidarity mechanism to be triggered off in the case of possible emergency situations regarding oil supplies. The petroleum crisis has meant to date a rise in prices rather than a physical scarcity of crude oil and happily, therefore, there has been no need to test the IEA's power of action with regard to the distribution of oil stocks under emergency conditions.

The Italian Energy Outlook to the Year 2000

What Italians see as Italy's future domestic and international role determines its energy-policy objectives. If that role is to be competitive in world markets, then certain highly specific conclusions emerge regarding energy saving, energy consumption, and the long-term, short-term, and medium-term energy outlooks.

Energy Saving

Containment of energy consumption is called for in particular. This must not be seen in terms of a reduction in the consumption of energy that would lead to a reduction in the quantity and quality of goods and services produced. Rather, a policy of energy-saving is needed that calls for avoiding waste; producing the same amount of goods and services with a smaller energy input (made possible by technological innovations that enhance the efficiency of the production process); and meeting demand with different goods and services requiring a smaller energy input.

Waste avoidance is possible in all branches of industry, and this is especially true of waste associated with heavy energy inputs. Statistics for 1979 show that the consumption increase of 3.5 percent for industry in 1979 was markedly lower than the 6 percent rise in the industrial production index. This confirms the trend of reduced consumption-per-unit-produced that began following the first significant price increase for oil. Bearing in mind that this phenomenon developed over a relatively short time span, one may ascribe it more to conservation measures than to effective technological innovation or to changes in the industrial product mix.

In view of technological innovations in industry, the potential for energy saving in those sectors with the highest energy inputs (steel, cement, paper, glass, and basic chemicals) will probably not be great. This is because there is generally an excess of productive capacity that does not justify the provision of new technologically more advanced installations. Also, many existing plants are relatively new, with high levels of energy efficiency. One area where action could take place is light industry. Here a well-conceived policy of demonstration and financial incentives, particularly for small and medium-sized forms, could produce as much as a 20-percent saving in energy.

In the private-transport sector, despite the long-standing tradition in Italy of small, low-weight vehicles, further energy-saving improvements are possible. These include adopting electronics in automotive applications, streamlining automobile bodies, and making more efficient engines. The savings that could be achieved by these means are estimated at 20 percent to 30 percent for heavy-duty vehicles and 30 percent to 40 percent for cars. However, these potential economies will be possible only with new models, which will gradually increase their proportion of the total number of vehicles on the road.

Other fields for action are total-energy systems based on the simultaneous production of electric energy and heat (cogeneration). Systems of this kind must be developed to fulfill the requirement for specific applications that can be integrated with other applications, that is, systems that can serve industrial and community needs. Here it is a question of evaluating the necessary outlay, not only in terms of energy but also in terms of the duration of payments, which in most cases will be protracted. In the domestic sector, active and passive heating optimization and temperature regulation could offer a field for immediate action.

These various approaches to energy saving will need to be backed up over the long run by other measures designed to act on the industrial mix. In addition to supplying energy demand, actions designed to influence the industrial mix must have as their principal goal the strengthening of a particularly fragile economy based on sectors that use mature, often highly energy-intensive technologies, and on sectors that use less mature, often less-energy-intensive technologies. The latter often produce consumer goods that compete with those of the developing countries.

Italian Energy Consumption to the Year 2000

The various approaches to the avoidance of waste, the development of new technologies, and the changes in the industrial mix, taken together, should reduce consumption of energy per capita. However, given the particularly long technical time scale for these solutions to take effect, it will be necessary, if the intention is to maintain a more-than-negligible economic growth, for energy-consumption levels to rise by an appropriate amount.

At present levels of per-capita energy consumption, the EXXON company has calculated that it should be possible by the year 2000 to achieve energy savings per unit of GNP on the order of 30 percent in the United States and 20 percent in Europe, with 1970 as the base year. For Western Europe, the study anticipates that energy demand will increase by a rate of 2 percent per year between 1980 and 1990 and at a rate of .7 percent per year between 1990 and 2000.[2]

In view of Italy's low per-capita levels of energy consumption and her general situation, allowance should be made for energy consumption to increase by a greater extent than in the other major industrialized countries. Nevertheless, the present high degree of dependence on hydrocarbon fuels, the domestic unavailability of substitute fuels, the difficulties in moving the nuclear program forward, and the long time span before renewable sources achieve a sufficient degree of maturity warrant the assumption of an especially austere scenario, with an average annual growth rate of 1.6 percent until the year 2000. In absolute terms this means passing from the present annual average consumption of energy of about 148 mtoe to about 203 mtoe by the year 2000. To achieve these targets assumes the operation of a severe energy-saving policy, with all that such a policy might imply in terms of penalizing income and economic development (see figure 10-1).

The Long-Term Role of Primary-Energy Sources

We have considered forecasts that can realistically be made about the future pattern of Italy's energy consumption. The picture needs to be completed, however, by indicating the forms of energy to which Italy should resort in order to strengthen the weaknesses of the present energy system and to meet new energy demands.

Electricity is increasingly in demand for those production activities in which technologies are most advanced—activities that are expected to gradually replace the basic industries that today are the heaviest users of primary energy. Here it may be noted that the patterns of overall electric consumption and economic growth can to some degree be disassociated from one another. Nevertheless, the pattern of electric-energy consumption is closely linked with the dynamics of economic and social development, and will therefore increase. Furthermore, conservation measures taken in the electrical-energy sector will be of negligible efficacy, given the high degree of efficiency of an electricity-based system.

Electricity has many characteristics that make it a mandatory tool for attaining more progressive increments in productivity and for attaining improved social conditions. These include its capacity to use all primary-energy resources; its ability to be distributed down to the capillary level; and its cleanliness, regularity, and reliability in operation. Electricity, to summarize, is an essential component if the emphasis on automation demanded for an effective energy-saving policy is to succeed. Accordingly, for the future as for the past, the electricity-demand elasticity values will continue to be inelastic and less susceptible to alteration than those values relating to energy consumption as a whole. Thus any assumption as to the contribution to be expected in the future from various primary-energy sources toward meeting energy demand must be consistent with the scenario of austere

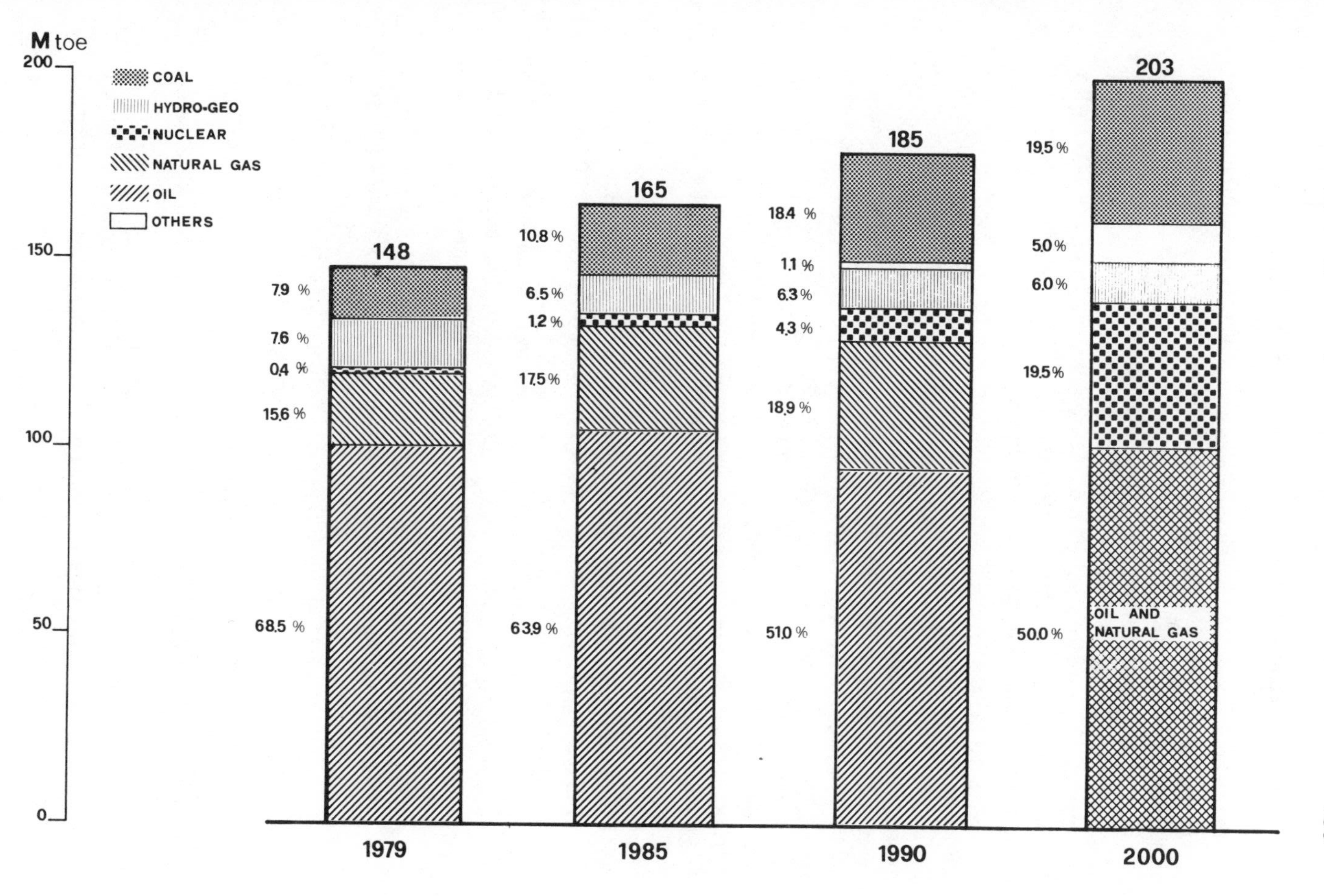

Figure 10-1. Italy: Forecasts of Energy Consumption

overall consumption, the need to develop economies of scale in industry, and the need for diversified and assured energy supplies.

The view is taken that a convincing energy policy must call for a reduction in the role of hydrocarbon fuels, so that they will represent a maximum of 50 percent by the year 2000. These will still be imported fuels, to which one may add a small fraction of domestic production. Such an objective is what the other principal industrialized countries that do not possess significant energy resources of their own expect to achieve. Thus, by 2000 hydrocarbons should represent 50-55 percent of the energy budget of Japan and 46-51 percent of the energy budget of France.

To attain Italy's objective it will be necessary to develop the use of coal considerably, to the point where consumption would rise to 60 million tons by the year 2000. This would satisfy approximately 20 percent of overall demand. The remaining 30 percent would be met from other sources that can be called domestic (nuclear, hydroelectric, geothermal, and solar energy). By such means, energy dependence would be reduced under the most favorable hypothesis from the present 82.5 percent to 63 percent by the year 2000—still a very high level when compared with the projections for the same date of 15-17 percent for the United States and 36 percent for Western Europe.

To attain the objective of 63 percent external dependence, the following will be required:

1. Maximum development of renewable sources of energy. These are negligible today but should come to provide a contribution on the order of 5 percent by the year 2000.
2. A three-and-one-half-fold increase in the contribution from coal. From approximately 17 million tons in 1979, this will increase to 60 million tons—practically all of this figure will go toward the production of electricity.
3. The placing of orders beginning in 1980 for two nuclear power plants of a 1000-MW rating each year, and their entry into service eight years from the date of the orders. This would mean 20 to 25 units in operation by the year 2000, including one at Caorso and two at Montalto di Castro.

Short-Term and Medium-Term Energy Outlook (1985-1990)

A number of actions, if started between now and 1990, could yield tangible results for Italy during this period. Other kinds of action might provide results in a later period.

As regards oil, it should be kept in mind that this energy source offers, and will continue to offer, the highest possible degree of freedom (within certain limits), though subject to all the political imponderables potentially influencing it. A greater use of oil entails purchase conditions that could prove extremely expensive.

In the natural-gas sector, it is taken for granted that option rights will be negotiated with Algeria for the import of additional amounts and for the doubling of the size of the pipeline now under construction. The present contract allows Italy to import 12,360,000 cubic feet of natural gas per year for twenty-five years. This will represent 42 percent of all supplies from this source by 1985. In order not to develop an excessive share of imports from the OPEC countries, import contracts will need to be negotiated with other countries. In particular, the Soviet Union could come to supply an increasing amount of natural gas, though for obvious security reasons only within the context of a comprehensive European energy policy.

In the coal sector, action can easily be taken (within limits) offering noteworthy results in terms of energy substitution and diversification. There are a number of power stations for which it is possible to replace oil with coal. According to the most recent ENEL estimates it should be possible even in the short run to convert from oil to coal at Vado Ligure, La Spezia, Brindisi, and Sulcis, thus saving 6.5 mtoe by 1990. Further coal-fired stations could be constructed. In its latest ten-year plan, ENEL proposes to build eighteen such plants by 1990, ranging in size from 300 MW to 640 MW.

If the programs were to be implemented in their entirety, it would be possible, according to ENEL, by 1990 to achieve an electric-power production from solid and derived gaseous fuels of about 22.4 mtoe as compared with 2.2 mtoe in 1979. A point to be emphasized is that coal supplies pose no problem of availability, considering their balanced geographic distribution. Over 60 percent of world coal reserves are found in countries notable for their political stability—the United States (27.9 percent), the Soviet Union (17.3 percent), the United Kingdom (7.2 percent), West Germany (5.4 percent), and Australia (4.3 percent).

Despite coal's advantages, there are drawbacks, including the difficult technical problems of transportation, combustion, and disposal of enormous quantities of ash. Coal power also faces serious economic (cost-of-plant), ecological, and climatic constraints. In addition, the siting of coal plants will affect local interests and may lead to further problems. The initial reaction of regional authorities must be considered in making site decisions. A final problem is that coal until recently has been used largely in the areas where it is produced. Coal trade, expanding in order to meet the demand of countries dependent on others for this fuel, will entail having to deal with particularly difficult infrastructural problems.

It is important to note that 34 mtoe of coal are projected for use in Italy by 1990. With 22 mtoe projected for electrical-energy production, approximately 12 mtoe is projected for nonelectrical uses such as steel, cement, or brickmaking; as well as domestic applications. However, there are serious imponderables concerning electricity-production using coal, given the difficulties attending the increase in only five years from 6 mtoe in 1985 to 22 mtoe in 1990. The total coal consumption in 1990 could reach a level considerably below even that given here. Should this be the case, it would be necessary to resort to a greater use of oil, to even maintain the reduced development model presented in this chapter.

One final comment about coal: the conditions will probably not arise in Italy to permit any substantial development of programs along the lines of those going forward or projected in the United States, West Germany, South Africa, or Australia for coal liquefaction or underground gasification. These are technologies that make sense especially where coal reserves are abundant. Only Japan, relying on her demonstrated economic, technological, and political strength, has any serious intention of developing programs of this kind, despite its lack of indigenous coal resources.

In the hydroelectric and geothermal generating sectors, residual resources could be harnessed but these, with the considerable investment and technological effort required, can make no more than a negligible contribution. In particular, the additional hydroelectric supplies that are technically feasible, irrespective of economic and environmental problems, might by 1990 amount to about 2.4 mtoe above the 10.7 mtoe provided by this source in 1979. Geothermal energy may be increased from the presently produced 0.6 mtoe to about 1.0 mtoe by 1990. On the whole, therefore, granted that the intensive R&D activity planned for the respective sectors proves successful, the contribution from hydroelectric and geothermal sources could increase over the 1979-1980 period from 11.3 mtoe in 1979 to about 14.1 mtoe by 1990, representing an increase of 2.8 mtoe overall. This is attended by many uncertainties and is in any case not extremely significant when compared with the overall increase in electrical-energy production for the 1979-1980 period of approximately 47 mtoe.

Nuclear energy is available today. The associated technology is as mature a technology as that required for producing electricity from coal. In all the industrialized countries, with the exception of Denmark and the Netherlands, nuclear energy has been accepted for some time now, despite widening social debate on its most pointed aspects, safety, ecology, and proliferation.

The example of the United States should be a lesson to all. The Three-Mile Island accident resulted in only a minor slowdown in activity connected with the operation of these power stations. In Italy, however, Three-Mile Island was one of the elements which, unfortunately, lent undue weight to an irrationally emotional reaction. Yet in the course of 1979 the

United States produced about 280,000 million kilowatt-hours (mkWh) of nuclear electricity, an amount equal to one-and-one-half times Italy's whole electric-energy production for the same year. About 70 percent of this amount was produced *after* the Harrisburg accident, a circumstance that only goes to show the passing nature of the event and the emphasis still placed on nuclear energy.

The potential decision to go forward with the Italian nuclear-power program is one that needs to be made immediately. It should not be forgotten that the need for an adequate evaluation both at the local and at the regional levels means long time delays before any such decision can be translated into concrete contributions to the energy system. Orders should be placed at the rate of two units per annum to help remove the episodic character so far attached to nuclear power and to ensure a minimum threshold of nuclear-plant commissioning, justified both by the survival of industrial-production complexes under efficient and competitive terms and by the maintenance of an adequate licensing and control apparatus.

The objective set for nuclear power plants is compatible with the requirement that CNEN, ENEL, and the regional authorities conduct a field examination of the places indicated in CNEN's map of sites. The purpose is to arrive at choices of sites that are not high-handed impositions but the outcome of honest comparisons and evaluations. Once full information is assured and all aspects of the problem have been worked out with all interested parties, the decision to build a plant must be made at the political level, offsetting the drawbacks at the local level with foreseeable benefits at the same level and to the community at large.

Clearly, the nuclear-power program entails considerable capital outlay over and above that necessary for the purchase of oil or other resources. For example, the cost of a nuclear plant is approximately 1.3 to 1.7 times respectively that of a coal-fired station or an oil-fired plant of comparable power rating. The time taken before a nuclear power plant comes on line is also long—1.4 to 1.5 times longer than that of other plants. This in turn means that a certain amount of capital will remain unproductive for a longer period of time than with coal or other types of power stations.

It should be remembered that during the phase in which the plant is being built, capital expenditure in the nuclear sector relates almost exclusively to engineering, manufacturing, and site-preparation activities. All these go into making up Italy's GNP and thus represent a limited drain on the country's foreign-exchange holdings. The share of the total outlay going to license fees, specialized technical aspects, and the purchase abroad of components that cannot, or cannot justifiably, be manufactured in Italy is reckoned at only 15-20 percent of the overall cost of the generating station.

Once a nuclear power plant is operational, the relationship between the cost of fuel oil and coal needed to operate a conventional power station and the cost of nuclear fuel needed for a nuclear power plant is respectively about 5 to 1 and 3 to 1. A 1,000-MW nuclear power plant operating in Italy calls for an outlay of 50,000 million lire per year for fuel as compared with 240,000 million lire for an oil-fired station and 110,000 million lire for a coal-fired station. When all cost components have been taken into account, one kilowatt-hour produced by oil is about 2.7 times more expensive than one kilowatt-hour produced by nuclear power; and one kilowatt-hour produced by coal is about 1.5 times more expensive than one kilowatt-hour produced by nuclear power.

The greater economy possible with nuclear energy will have a leveling effect on the types of resources used to produce electricity, especially coal. This will have beneficial effects on Italy's balance of payments. It will be possible for Italy, within the limits determined by social exigencies and by the progressive evolution of the industrial mix, to maintain its high value-added activities (in the cement, glass, and aluminum industries) which otherwise would shift to countries where conditions are more favorable. This will produce a positive though indirect effect on employment opportunities together with the direct and positive effect represented by the construction of generating stations and fuel-cycle plants.

In conclusion, the difference between nuclear and other sources of energy employed today for producing electricity lies in the fact that the use of nuclear energy saves in the cost of energy and creates employment opportunities and economic activities, contributes to improving the balance of payments, uses to good purpose the national labor force in a spearhead sector, and creates benefits where international markets are concerned.

The results one expects to obtain must of course be placed in the context of the long-run energy strategy with the goal of achieving a major reduction in the role of hydrocarbon fuels. The forgoing analysis points to the lines of action which, once a start is made and some stage of completion is reached, can lead to a significant reduction in Italian dependence on hydrocarbons and to a reduction, though not a major one, in energy dependence.

Impact of Energy Policy on the Economy and Society

The energy strategy illustrated in this chapter is based on the resort to many oil-substitute energy sources and on a decisive resolve to conserve energy. Hitherto the costs of obtaining energy supplies have not had a determining impact on the economies of the industrialized nations; but there is every presumption that energy will come to weigh much more heavily in future decades.

In the industrialized countries, the energy replacements of the past have had a marked structural impact on the productive systems and on the associated population settlements. The change from renewable and dispersed sources of energy such as human labor, water, and wind to nonrenewable wood and coal which came about contemporaneously with the Industrial Revolution of the last century encouraged a progressive centralization of production and the emergence of a more centralized urban society. An example of this was the development of urban centers near coal deposits in nineteenth-century Great Britain.

When coal was replaced by oil—a process that began in the first half of the present century—another centralizing effect resulted, this time because of the growing economies of scale associated with industrialized production as well as the development of the transport sector. Where this has taken place (and the process has been favored by a continuing mechanization with a consequent increase in productivity), a gradual abandonment of the countryside and a corresponding formation of megalopoli resulted, swallowing up existing suburbs in the process.

A rising consumption of natural gas represents in one sense a further step toward centralization. In the future, only those served by city gas services will be able to benefit from the expanded role of this source. As a reciprocal process, the natural-gas distribution system demands the provision of a capital-intensive infrastructure whose economies of scale will be all the greater as the built-up areas expand.

The progressive electrification of our society, such as has taken place up to the present does indeed make available energy of a kind offering ease of transportation and distribution over wide areas; but it has also helped in the centralization process. For electricity has been widely employed by industry and in civil applications by towns, and the installations producing it are usually enormous and complex undertakings, highly capital-intensive. In this sense, obtaining electricity from nuclear sources is an even-greater centralizing force, though it should be remembered that electric energy itself could actually promote a decentralization process.

The need to be able to exploit alternative renewable resources over long periods of time—solar energy in particular—demands that more careful attention should be paid to the decentralization process. Indeed, as has been amply demonstrated in a recent study made on behalf of the European Economic Community, the optimal use of solar energy in countries such as Italy calls for geographically decentralized establishments, because the solar source is diffuse and distributed throughout the national territory.[3] Again, it is abundantly clear that however many incentives may be proposed in order to promote the penetration of renewable forms of energy, these will only be obtainable at the industrial level well into the twenty-first century.

Italy is endowed with innumerable small towns possessing notable historical and cultural values, craftsmanly traditions, and a lively entrepreneurial vocation. Accordingly, Italy must take care not to waste this heritage on a centralizing process of which the present pattern of the energy system is but one example. However, difficulties would arise if, once this phase were over, it were proposed to massively promote the penetration of alternative sources of energy by means of a decentralizing action.

In regard to the energy system, one arrives at a conclusion similar to the conclusion that may be drawn from analysis of the industrial sector as a whole. However much it is attempted to promote a model based on a multiplicity of small-sized businesses with a high entrepreneurial coefficient and a high degree of specialization, it is unthinkable that Italy could fail to accord an equally fundamental role to the large firms that constitute one of the main pillars of its industrial society. The pattern of our industrial structure, with its plurality of roles and dimensions (faithfully mirrored in the energy sector as well), is the solution to be sought. But it must be pursued with a clarity of vision, avoiding the pitfall of replacing the strategic approach with a succession of uncoordinated tactical approaches in conflicting directions.

Notes

1. See, for example, Ministry of Industry, *Piano per l'Energia* (Rome: 1981).
2. Exxon Corporation, *World Energy Outlook*, 1980, 1981.
3. Umberto Colombo, and Oliviero Bernardini, "A Low Energy Growth Scenario (2030) for the Europe of the Nine," Working Paper prepared to assist the panel report *In Favor of an Energy-Efficient Society*, prepared for the Commission of the European Communities (Brussels: Collection Studies, Energy Series no. 4, 1979).

11 Energy Policy in the European Communities

Wilfrid L. Kohl

The European Communities (EC) are viewed by their proponents as "the first stage of an economic federation."[1] They have some federal features, for example, the common agricultural and commercial policies. Gradually, and with difficulty, they are trying to acquire broader central powers. Two energy sectors were part of their original mandate: coal was addressed early in the ECSC; and nuclear cooperation was tried in Euratom. Both sectors were soon beset with major problems. The growing share of cheap oil in the energy market caused a major contraction of coal production. Nuclear cooperation was slowed by France, which was not interested in merging her advanced programs in a common European effort, and by other difficulties including government-industry relations.[2] The formation of the EEC, or Common Market, brought about the need to shape common policies in other economic sectors. Yet, according to the Treaty of Rome, these policies were only to be coordinated. Thus, the EC do not have a full mandate to direct the formation of a comprehensive energy policy; and because of cheap oil there was no need for such a policy until the oil shock of 1973-1974.

The first oil crisis revealed the vulnerability of the Communities, which had become heavily dependent on oil imports, and thus underscored the need for a common European energy policy. But early efforts after the crisis to forge such a policy were limited to collecting data on Europe's energy situation, setting objectives, making recommendations to member governments, and supporting some research and demonstration projects. The second oil shock provided further impetus. Recently the Commission of the EC has sent a stream of more ambitious energy proposals to the Council, but only some of these proposals have been approved.

The Commission's approach is as follows: it "does not claim that everything in the energy sector should be regulated on a centralized basis at

---

The European Communities (EC) is the collective designation of three organizations: the European Coal and Steel Community (ECSC), formed in 1952; the European Economic Community (EEC or Common Market), formed in 1958; and the European Atomic Energy Community (Euratom), also formed in 1958. The executive bodies of the three Communities merged to form a unified Commission in 1967. But the Commission is still charged with implementing the three original treaties. In addition to the Commission, the institutional structure of the EC includes a Council of Ministers, a Court of Justice, and the European Parliament—now directly elected. The ten full members of the EC are Belgium, Denmark, France, West Germany, Greece, Ireland, Italy, Luxemburg, the Netherlands, and the United Kingdom. In 1977 Portugal and Spain applied for membership.

Community level. A large part of the strategy on which the Member states have agreed can be implemented only at national level. But . . . these measures must be coordinated and, where necessary, supplemented and reinforced by Community measures."[3] The philosophy underlying the EC's approach to energy policy calls for both national and international intervention in the market in order to ameliorate Western Europe's situation of substantial dependence on energy imports. Although progress has been slower than hoped for in forging a common energy policy, the Communities possess some institutional advantages in contrast to the International Energy Agency (IEA): first, EC decision-making is a legislative process and decisions are binding on member governments; second, the Communities possess a larger budget and can support, for example, RD&D projects (usually limited to partial support); and third, France is a member of the EC.

The present state of EC energy policy can be summarized under the following headings:

Targets and guidelines.

Regulations in the field of energy saving.

The development of nuclear energy.

Support of research, development, and demonstration (RD&D).

An emergency oil-sharing system.

Energy diplomacy.

Targets and Guidelines

Following the first oil crisis, the primary concern was to reduce oil consumption, which amounted in 1973 to about 61 percent of total EC energy consumption. To do so, it was necessary to reshape EC energy policies. Objectives were set for 1985 as shown in table 11-1. When it soon became clear that the objectives based on the hope of reducing oil to 40 percent of energy consumption by 1985 were too optimistic, the Community concentrated on the 50 percent objective.

How did the EC perform? By 1979, oil had been reduced to about 55 percent of energy consumption.[4] Progress has certainly been made, because of slower economic growth, fuel switching, and conservation. The development of North Sea oil, which now covers about 16 percent of EC oil demand, has expanded indigenous EC production. Natural gas has steadily increased its share in total primary-energy consumption, supporting about 18 percent of total requirements in 1979. Thus, the 1985 target has already been reached in this sector. The gas share could increase to 20 percent in 1985, though gas imports will have to cover an increasing part of consumption.

Table 11-1
European Communities' Primary-Energy Objectives for 1985 (1974 Objectives)
(percentages)

	For the Record		*1985 Objectives*	
Type of Energy	*1973 Estimates*	*1985 Initial Forecasts*	*50 Percent Dependence*	*40 Percent Dependence*
Solid fuels	22.6	10.0	17.0	17.0
Oil	61.4	64.0	49.0	41.0
Natural gas	11.6	15.0	18.0	23.0
Hydroelectric	3.0	2.0	3.0	3.0
Nuclear energy	1.4	9.0	13.0	16.0
Total requirements	100.0	100.0	100.0	100.0

Source: *Official Journal of the European Communities,* Council Resolution of December 19, 1974, C153/3 (Brussels: 1974).

Current gas imports are about 25 percent, a figure that may climb to 50 percent by 1990. The coal share of total EC energy requirements was about 20 percent in 1979 (roughly comparable to that of gas), thus indicating achievement of the initial objective in this sector also. Moreover, the coal sector is ripe for further expansion and is expected to double by the end of the century, though most of the increase will have to come from expanded imports.

Nuclear energy, which in 1979 contributed only 3.2 percent of the Communties' energy needs (11 percent of electricity production) is the biggest disappointment, far from the target set in the 1974 objectives. Hydroelectric, geothermal and miscellaneous forms of energy contributed about 3.5 percent in 1979.

In June 1980 the Council of Ministers, spurred by the impact of the oil-price shock of 1979, set these new guidelines for 1990:

The average ratio of the rate of growth of gross primary-energy consumption to the rate of growth in GDP (Gross Domestic Product) is to be reduced to 0.7 or less.

Oil consumption is to be diminished to about 40 percent of gross primary-energy consumption.

In the area of electricity production, 70 percent to 75 percent of requirements are to be met with solid fuels and nuclear power.

The contribution of renewable energy sources is to be encouraged and expanded.

An energy pricing policy is to be established.

In its decision, the Council attached priority to progress in energy saving, the rational use of energy, and the reduction of oil consumption and imports. In order to strengthen the EC's ability to monitor member-state energy policies and encourage their convergence, the EC Commission will henceforth request its members to submit annually their energy programs up to 1990. It will review national programs and submit an annual report to the Council (a process somewhat analogous to the IEA annual review activity). The Commission is charged to "make recommendations and proposals with a view to increasing the convergence of the Member States' energy policies, ensuring that (EC) energy objectives are achieved and adapting them to long-term economic trends and energy supply conditions."[5]

In 1980 there was a sharp decline of 4.6 percent in overall EC energy consumption. Although in some respects 1980 may have been an abnormal year, the statistics show that the link between economic growth and energy consumption is being broken. Average GDP growth was 1.3 percent. Decreases in gross inland energy consumption ranged from 1.2 percent for Italy to 9.2 percent for the United Kingdom. Factors explaining the significant fall in consumption include the continuing poor economic situation (especially in the steel industry, a major energy-consuming sector), the impact of the oil-price spiral in 1979, overall higher temperatures compared with the previous year, and the initial effects of new measures aimed at more efficient energy use. The tendency of oil's share to decrease in overall energy consumption in favor of coal and nuclear power, a trend begun in 1979, was confirmed, as the oil share fell from 54.2 percent in 1979 to 51.8 percent in 1980. Other figures are shown in table 11-2.

Besides the impact of higher oil prices and general economic conditions, such progress as the Communities have made in reshaping their energy policies has, of course, resulted mostly from the efforts of the member states, where sovereignty in energy policy mainly lies. However, the Commission has played a useful role in prodding the countries into action.

Conservation/Energy Saving

In another resolution in June 1980, the Council attempted to step up the EC action in the field of energy saving. It was agreed that each member state should have by the end of 1980 "an energy saving program covering all the main sectors of energy use and an appropriate energy pricing policy."[6] Guidelines were recommended to member states on energy pricing policy (essentially that consumer prices should reflect conditions on the world market). Regulations have also been recommended for energy saving in the home, in industry, in agriculture, in commercial offices, and in transport.

Table 11-2
Gross Inland Energy Consumption for the Nine Countries of the European Communities

Type of Energy	*1980*		*1979*		*1980/1979*
	mtoe[a]	*Percentage*	*mtoe*[a]	*Percentage*	*Percentage Ratio*
Hard coal	191.7	20.7	190.8	19.7	+ 0.5
Lignite (and peat)	28.8	3.1	28.9	3.0	− 0.3
Crude oil	479.6	51.8	525.2	54.2	− 8.7
Natural gas	168.1	18.2	172.5	17.8	− 2.5
Nuclear power	42.6	4.6	37.2	3.8	+14.5
Primary electricity and other	15.0	1.6	15.1	1.6	− 0.7
Totals	925.9	100.0	969.7	100.0	− 4.5

Source: Eurostat, Statistical Telegram (Luxembourg: April 15, 1981) p. 2.
[a]Millions of metric tons of oil equivalent.

The Council requested the Commission to monitor national energy-saving policies and report to it on progress made toward attaining Community energy-saving targets. Subsequently, a first inventory has been prepared by the Commission, prior to more analytical work and recommendations to member states.[7]

However, it should be noted that the EC's actions on energy conservation and saving have for the most part been limited to guidelines and recommendations, leaving implementation to the member states, the companies that produce various appliances and other items, or to the consumer, according to national priorities and conditions. Although in 1979 the Council did adopt directives requiring the labeling of certain household appliances and electric ovens according to energy consumption, other Commission proposals in 1980 extending the required labeling to other high-energy-consumption appliances such as washing machines, dishwashers, refrigerators, and freezers have been left in the status of recommendations by the Council.

The Development of Nuclear Energy, Safeguards, and Nuclear Supply

According to official policy, the expanded use of nuclear energy, along with greater use of coal and energy saving, offer the keys to reducing significantly EC dependence on imported oil. Nuclear power offers a production-cost advantage compared with coal and oil for base-load electricity generation. The Communities have pursued a policy favorable to reprocessing of nuclear fuels and the development of fast-breeder reactors, given the low level of indigenous European uranium resources.[8]

Yet nuclear power only accounts for 4.6 percent of EC energy needs today. Several member-country nuclear programs have encountered serious delays because of domestic opposition. International uncertainties related to insecurity of nuclear supply and nonproliferation have also been an inhibiting factor. At the policy level, therefore, the Communities have not been able to achieve a consensus to promote nuclear power actively in a major way.

There have been a few low-level efforts. Aside from the research and exploration activities already alluded to, the Commission has held public hearings (1977 and 1979) to enhance public awareness of the facts relating to nuclear power development. A Euratom loan facility was set up in 1977 to facilitate nuclear investments in power stations and fuel-cycle facilities. About 1,000 million European Units of Account (1EUA = $1.23 in February 1981) was to have been committed to power-station projects in various EC countries by the end of 1980, and a proposal was being prepared by the Commission to extend the program.

To implement the Euratom Treaty of 1957, the Commission maintains Directorates for Nuclear Energy and Euratom Safeguards (under Energy); and Research, Development, and Nuclear Policy (under the General Directorate for Research, Science, and Education). There are also the Euratom Supply Agency and four nuclear-research establishments working under the Joint Research Center.

A major continuing Euratom activity under Chapter VII of the Treaty is in the area of safeguards. Indeed, Euratom instituted the first multilateral safeguards system. The Treaty provides for control and safeguards over all fissile material in the Community, except for material used for national defense purposes. Euratom inspectors continue their work today, operating from the Safeguards Directorate located in Luxembourg. A major complication was introduced by the Nuclear Non-Proliferation Treaty (NPT), effective in 1971, which designated the International Atomic Energy Agency (IAEA) as the principal international agency to implement safeguards. Following complicated negotiations, the three EC agreements were concluded with the IAEA that allow for joint teams made up of inspectors from both organizations to inspect nuclear reactors in European countries in order to meet conditions of the NPT. In this context, the continuing safeguards role of Euratom seems redundant, but the compromise was apparently necessary in order to preserve the Community's historic role in this area.

The Euratom Treaty provided for a common-supply policy based on the principles of equal access to supply. Contracts were to be negotiated for the supply of ores from within and outside the Community by the Supply Agency. Since the Community depends on imports for 80 percent of its nuclear fuel, recent supply instabilities, such as the shift in American policy under the 1978 Nuclear Non-Proliferation Act, have caused considerable concern. The Euratom/U.S.A. nuclear cooperation must be renegotiated. Euratom also has cooperation agreements regarding nuclear supply with Canada and Australia. In fact, however, there are also bilateral agreements between some EC countries and major suppliers, especially because of the interface with military nuclear cooperation.

Under Article 70 of the Euratom Treaty, the Community makes grants to assist exploration for new uranium supplies. New resources have recently been identified in Italy and Greenland. Of the EC countries, France is the only one with secure access to major uranium deposits.

Energy Investment: Research, Development, and Demonstration (RD&D)

Unlike the IEA, there is an EC budgetary mechanism for contributing to energy RD&D projects. The energy research program costs about $400 million annually and makes up about 70 percent of the total EC RD&D funds. This is equivalent to about 10 percent of the total amount spent by member governments in a year on national RD&D programs. EC loans to the energy sector have been increasing, representing 46 percent of total loans in 1979 and 41 percent in 1980.

The RD&D program of the Communities has three main objectives: the development of nuclear technology—especially nuclear safety and waste-disposal techniques; energy-saving demonstration projects; and development of new energy sources—especially nuclear fusion, solar energy, and geothermal energy.

Initially, Euratom sought to promote nuclear research in the member states and to complement it by organizing Community-level research and training programs. Research is pursued through the Joint Research Center (JRC). Currently, priority is given to research on reactor safety and waste-disposal problems, with the aim of helping to reduce public doubts about the safety of nuclear power programs. EC R&D expenditures have been augmented recently, especially for reactor safety, in the 1980-1983 budget allocation. The Communities spent approximately 40 percent of their RD&D budget in 1977-1980 (346 million EUA) for nuclear-safety-related research.

Though R&D efforts were traditionally focused on the nuclear field, more recent attention has been paid to alternative-energy sources and to conservation. A second four-year R&D program (indirect action only) is under way, covering the period 1979-1983, with funding at the level of 105 million EUA.

The general attitude toward R&D is as follows: The Communities will support a few very expensive projects, such as the Joint European Torus (JET) project, that no country could afford on its own. They can pool knowledge in other areas that no single country possesses and promote joint advances, such as in the case of the one MW solar power plant being supported in Sicily. The EC can look at longer-term energy needs, and prod countries into action in these areas, while holding out some financial incentive and playing a coordinating role. Sometimes the Communities can break new paths in research, though usually they pay no more than 50 percent of the cost of projects. Governments (or sometimes private industry) pay the rest.[9]

By far the most expensive program is the Joint European Torus (JET) project established in 1958, which is meant to place the EC at the international forefront of controlled-thermonuclear-fusion research. The Communities finance 80 percent of the project, which is located at the Calham Research Laboratory in the United Kingdom. Construction costs have been estimated at 200 million EUA. Other EC activities are organized through the direct-action program under the Joint Research Center, or the indirect- and concerted-action programs carried out in association with member governments.

To demonstrate the industrial and commercial viability of alternative energy sources, the Council launched a scheme in 1978-1979 of granting support to demonstration projects.[10] Support may not exceed 40 percent of the cost of a project, and half the support is repayable by the recipient if the project yields methods that are successfully marketed for industrial or commercial purposes. In 1979, 95 million EUA were allotted for a five-year program, divided as follows: 50 million EUA for coal liquefaction and gasification, 22.5 million EUA for geothermal energy, and 22.5 million EUA for solar energy. In response to a first-round invitation for bids, 183 requests were received, of which 43 projects were financed at a level of 31 million EUA. A second-round invitation was issued in 1980. The interest aroused has led the Commission to ask the Council to raise the amount of support for liquefaction and gasification projects from 50 to 100 milion EUA. The Commission also plans to ask for more support for solar energy and geothermal projects. However, the method used by the Commission has some flaws: it simply asks for bids, and does not itself get involved enough with the industries conducting the projects.

A similar approach is being followed in the area of energy conservation, where demonstration projects have also been given partial support. In 1979 55 million EUA were allotted for a four-year program. Some 53 projects were selected in a first-round invitation, costing 21.4 million EUA, out of 324 applications. Another round of invitations is under way. The Commission also plans to ask the Council for more funds in this area, since the amount allocated for the four-year program clearly will not be sufficient.

The amount of funds available to the Community for investment in energy projects is still relatively modest, even if the additional facility for loans from the European Investment Bank is taken into account. To meet this situation, the Commission proposed a new energy initiative to the Council in March 1980. It called for additional funds on the order of 50,000 to 100,000 million EUA to augment the estimated 400,000 million EUA that member states are planning to invest in energy between 1980 and 1990. Various formulas were suggested for raising this money, including the idea that a tax could be imposed on Community energy consumption, production, or imports. This was a controversial proposal, and the Council did not act on it. Etienne Davignon, the EC energy commissioner who took office in January 1981, is reported to favor a less ambitious approach, supporting energy investment through loans and guarantees. In general, he seems to prefer to rely more on national-energy and investment policies, with the EC encouraging cooperation and coordination where possible.

Oil-Crisis Management: Emergency Sharing and Other Measures

In the 1970s an EC system of emergency oil management was agreed on consisting of four kinds of measures: oil stockpiling, coordination to reduce the effects of supply difficulties, the reduction of demand, and a surveillance mechanism of intra-EC trade in oil products.[11] The arrangements are compatible with those of the IEA.

The stockpiling of oil was required in the Communities as early as 1968. Current regulations require the member states to maintain a minimum of ninety days' supply of oil or oil products. In case of a crisis there is a mechanism of consultation between member states and the Commission regarding the allocation of oil stocks, a mechanism which may be triggered as soon as the gap between oil supply and demand reaches 7 percent. Measures to reduce consumption are envisaged in phases: the first phase will bring about a 10 percent reduction in consumption of petroleum products; a second phase includes reallocation among member states of quantities saved; and a third phase to deal with a severe crisis, with reductions of consumption above 10 percent, and reallocations. Reallocation under the EC mechanism comes about in a different way than under IEA provisions, that is, by controlling the trade or exchange among the EC member states in oil and oil products. A surveillance mechanism of the oil market is provided to regulate product flows in a crisis, overseen by a special task force at the Commission.

Following the Iran-Iraq war, the Energy Council—reacting as usual to proposals from the Commission—considered measures at its meeting in November 1980 to discourage price increases in the oil markets and to control spot-market trading. The member states decided to ask oil companies to draw on stocks in excess of current reserve requirements rather than make spot-market purchases. The Commission and the member states will coordinate and verify that actions are taken. Supplies are to be adjusted so as to correct imbalances among the member states. The Council urged that members encourage the further saving of oil in the public and private sectors and the substitution of other forms of energy for oil. Domestic production should continue at a high level. In dealing with the subcrisis situation, "The main objective will be to avoid an overall demand for imported oil at a higher level than that which can be made available by the producing countries."[12]

Moreover, the Commission is continuing to monitor trends in supply and demand and in stocks. It has recently conducted a study of the structure and operation of the spot market in cooperation with industry, including a register of spot transactions. Steps are being taken to establish the capacity to reintroduce a more detailed register in case of future disruptions on the oil market, in order to generate data and improve market transparency. There is also talk of a "code of conduct," which buyers and sellers on the spot market might be willing to accept in order to maintain the confidence of governments in their trading practices.[13] In its June 1981 meeting the Energy Council, led by Mr. Davignon, indicated a willingness to move ahead more quickly than the IEA, if necessary, to consider precise proposals for security oil stocks and their management in a subcrisis situation. Further proposals are expected from the Commission.

Domestic Constraints

A major cause of the difficulty in extending EC energy policy lies in differences in domestic economic structures of the member countries. Policy differences also develop out of the ideologies of parties in power and certain influential political groups regarding intervention in the market.

In terms of energy-resource endowments, the EC countries fall roughly into three groups. The energy-rich countries, which tend to resist EC energy policies, include the United Kingdom (with its North Sea oil and gas reserves and considerable coal deposits), and the Netherlands (which possesses rich offshore gas in the Groningen fields). The Dutch decided recently not to renew their gas-export contracts in order to preserve their resources. Britain is expected to attain net self-sufficiency in oil between 1981 and 1988. The British government wishes to retain sovereignty and strict control over the North Sea fields, but substantial amounts of British oil are now being exported to the Communities, equivalent to about 16 percent of total EC oil imports. Yet the British have resisted formal commitment within an EC energy policy to increase these exports, or to share more oil in a crisis. The United Kingdom is continuing a substantial nuclear program, and cooperates with Germany and the Netherlands on development of the fast-breeder reactor.

A second group of energy-poor countries—Denmark, Italy, and Ireland—have the most to gain from stronger energy policies in the Communities. Italy depends on imports for 80 percent of its energy and imports 99 percent of its oil. The country has small deposits of natural gas and has developed hydroelectric power, but it has no coal and its nuclear program is stalled. Moreover, the Italians have an excess of capacity in their petrochemical industry in Southern Italy for which they are constantly seeking EC aid. Denmark and Ireland have few domestic energy resources and must depend on conservation and policies of diversifying imported energy sources to reduce their vulnerability, though there is limited scope for exploring and developing small amounts of domestic or offshore oil and gas.

In the middle group of countries with respect to energy resources are France, West Germany, and probably Belgium. The countries in this middle group tend to favor some proposals for EC policy but not others, depending on the energy source involved. France's energy strength lies in its nuclear program, scheduled to provide about 30 percent of its energy needs in 1990. It also has some coal, gas, and uranium deposits, but is otherwise heavily dependent on imports for 75 percent of its energy needs. The French import 98 percent of their oil. The Federal Republic of Germany has large coal reserves of both cheap (but relatively low-quality) brown coal and very expensive deep-mined hard coal. The German nuclear program has been delayed because of domestic resistance. West Germany imports 58 percent of its total energy requirements and 96 percent of its oil. Belgium has some coal deposits, but with high production costs; and a substantial nuclear program that already produces over 25 percent of generated electricity.

Further complicating the situation are different economic structures and different national approaches to the role of the state in the economy. Following a free-market approach, the German government sets the framework in which the private sector operates. In England under the Thatcher government a similar philosphy reigns, though it inherited considerably weightier public enterprises that continue to give the state a substantial role in the energy market. France traditionally tends toward a dirigiste approach. The centralized French strate possesses multiple instruments to intervene in the market through a wide array of public enterprises in the energy sector, greatly facilitating the implementation of policy. Italy is located somewhere between the state and the free-market models.

Although there are several public energy enterprises, frequent changes in government inhibit the making of a stable energy policy.

How the wide disparity among the domestic structures of EC countries can inhibit EC decision making on energy was illustrated in the failure to reach agreement at the May 1978 Energy Council on aid to the refining and coal sectors. Italy had insisted for some time on a special aid system that would provide payment of a fixed EC subsidy per ton of suppressed distillation capacity, as compensation for the loss of economic activity from closure of refineries. The reason for the Italian intransigence was rooted in the particular situation of the Italian oil industry, primarily controlled by ENI, a state-owned company with few other possible sources of compensation. The United Kingdom and the Federal Republic of Germany, countries where the oil industry is more diversified, refused to accept the Italian position. As for aid to the coal industry, Italy was reluctant to go along with other countries partly out of a need to maintain this issue as bargaining leverage, and partly because she produces no coal domestically and therefore would receive no benefit under any new EC scheme. This is the perennial problem in the Communities of "juste retour."[14]

There are different degrees of consensus in the various European countries regarding the directions of energy policy, and different levels of politicization. State policies are subject to changing emphasis or directions as new parties or coalitions come to power. For example, under the present Tory government, the British have readjusted their energy policy to rely more on the free market, within certain institutional and political constraints. Since civil servants probably cannot be entirely relied on to carry out this policy, energy policy is made at the ministerial level in Cabinet. The most viable policy change is the rise in prices of oil and gas. Another shift is the elimination of majority equity interest of the British National Oil Corporation and the British Gas Corporation in North Sea oil and gas production. Energy Secretary David Howell has staunchly defended the government's energy-pricing policy against Labor attacks in Parliament, rejecting any suggestions of two-tier pricing. The approach is very compatible to that being developed by the Reagan Administration in the United States. But there are limits, of course, to what is politically possible. Opposition of the Trade Unions Congress (TUC) and of the Labor Party in Parliament has slowed Tory intentions of closing down unproductive coal mines. And it is unlikely that state energy agencies will be dissolved because of unprofitability.

In the already highly centralized French economy, the coming to power of the Socialists under President François Mitterrand can be expected to have less impact on energy policy. One possible shift, however, concerns the role of nuclear energy. The Mitterrand government initially halted the Plogoff reactor and placed a moratorium on new-reactor construction pending a full debate in Parliament on the role of nuclear energy. The attitude of the Miterrand government in general toward the institutions of the European Communities was not fully clear at this writing.

In Germany, the domestic political constraints on Chancellor Schmidt's professed intention to increase the role of nuclear energy seem formibable, constraints which could affect the German position in the EC and in other international bodies on nuclear policy. Both the Social Democratic Party (SDP) and the Free Democratic Party (FDP), which together form the governing coalition, have militant left wings that are against nuclear power (and increasingly anti-NATO as well). This is placing a serious constraint on both coalition leaders, Chancellor Schmidt and FDP Foreign Minister Hans Dietrich Genscher. Another kind of political issue sometimes creates tension within the governing coalition between the Chancellor and his Minister of Economics, Otto Lamsdorf, who also comes from the FDP. The SPD trends to favor increased government intervention in the energy sector, whereas the FDP is inclined to rely on the operation of the free market to work things out. This can reduce flexibility in EC or other international negotiations.

But sometimes the hard-nosed, aggressive Chancellor Schmidt has used international forums to support his views against opposing domestic forces. At the 1980 Venice Summit, for example, he was more supportive of nuclear energy than he had been in domestic meetings of the SPD party leadership. And at the 1979 Tokyo Summit, Chancellor Schmidt is reported to have overruled Minister Lamsdorf in order to clear the way for German agreement to oil import ceilings.

Energy Diplomacy: Relations with the IEA and the Euro-Arab Dialogue

Whatever the Communities do to improve their energy position, their heavy dependence on imports will continue. They need leverage in the outside world to guarantee security of energy supplies. All the EC countries except France are members of the IEA, and many of them place somewhat greater importance on the latter organization with its broader membership of almost all important industrial countries, including significant energy-producing countries such as the United States, Canada, and Australia. If a major cut-off of Europe's oil supplies from the Middle East were to occur, assistance would depend mainly on non-European IEA allies (in addition to some possible surge capacity from North Sea resources controlled by Great Britain and Norway). Moreover, the United States remains a crucial energy ally for Europe in view of advanced American technology in certain energy fields, and because the United States still plays the major security role in the Middle East (the source of much of Europe's oil), and has greater leverage on both sides of the Arab-Israeli conflict.

So far the IEA Secretariat has demonstrated stronger analytical capability on energy issues and has gathered more data on the oil market than has the EC Commission. Although it does not have legislative powers, the IEA has received high-level attention from member governments and its recommendations therefore carry a good deal of weight. EC energy measures, especially on emergency oil sharing, have thus become useful supplements to the IEA and are designed to be compatible with IEA measures. And the Communities do serve as an important link to France, which does not participate formally in the IEA.

In their efforts to diversify energy supplies, several EC governments and private companies were negotiating with the Soviet Union in 1980-1981 regarding construction of a pipeline from Siberia that would carry natural gas to Western Europe. Much of the equipment would be supplied by a European industrial and banking group, led by West German companies and banks. European coal-processing and other technology is reportedly being offered as part of an agreement in return for Soviet natural-gas supplies. It is noteworthy, however, that the Communities have not been involved in these negotiations.

The Communities have played a role in exploring increased North-South energy and related economic cooperation through the Euro-Arab dialogue, which has given the EC a more positive image vis-à-vis the Arab OPEC states than that enjoyed, for example, by the IEA. However, the results have so far been modest. Following the first oil shock, the Communities declared their readiness for a Euro-Arab dialogue, and discussions began in Cairo in June 1974. The dialogue continued periodically until 1978, when it was broken off by differences of view over the Camp David Accords. Following the Venice declaration of the European Council in June 1980, which announced Europe's willingness to explore a peace initiative in the Middle East that would include some role for the Palestinian Liberation Organization (PLO), further discussions ensued aimed at resuming the Euro-Arab dialogue, but political problems have continued to delay formal talks. The dialogue has established a certain institutional infrastructure: a General Commission and various working parties. Subjects discussed to date include agricultural and rural development, trade, investment, technology transfer, solar energy research, cultural cooperation, investment, and cooperation between development-aid organizations.[15]

Conclusion

> The Community, in the areas already transferred to it, has largely taken on federal features, for instance with respect to the Common Market and the Agricultural and Commercial Policy . . . In other areas, the Community

> has worked out cooperative forms which make it more like an association of states, for instance in economic policy in all its aspects. It may be said for the present that most of the fundamental and vital issues fall into this cooperative sector. In this fashion, a new type of conjunction of states is emerging which has no model in the past and has not yet taken on final shape.[16]

This characterization of the present state of European economic integration by a longtime participant-observer in European policy provides a useful background against which to view the place of energy within the EC framework. Energy policy is very closely related to industrial policy, which itself is a subset of general economic policy. While the Common Market (European Economic Community) brought about the need to establish common policies in the main sectors of general economic policy, according to the Treaty of Rome overall economic policies were only to be coordinated, not unified. Thus, the forging of common energy policies in the Communities is a painful, slow process of coordinating national energy policies. The process depends on striking compromises and building consensus. It is hindered by differing economic philosophies among the member states, by different resource endowments, and by economic structures that have been shaped by different historical development.

The Communities derive their legitimacy from the member states. The European Commission remains the only independent body, but it only proposes measures to the Council of Ministers, which represents member governments. The Council continues to make decisions by unanimity on all important matters. The preeminence of member-state sovereignty has been reinforced in recent years by the more regular meetings of the EC heads of state at the summit level in the European Council. Thus, in the end, the Communities are stuck with a tedious process of building consensus among the member states on the need for common policies, seizing opportunities for decisions as they arise.

Despite the difficulties, the European Communities perform a useful—but limited—role in energy policy. They serve as a clearinghouse and coordination center for national policies, occasionally prodding member states by setting targets and guidelines, and even regulations in a few sectors. While EC research-and-analysis capabilities are not as good as those of the IEA, the Communities are taking steps to improve their performance in this area and now plan regular reviews and appraisals of national energy programs. One of the areas in which the EC Commission is structurally best equipped is in financing RD&D, since it has a budget to contribute to projects. Its record so far in that area is controversial. Progress has been made recently in the areas of energy saving and oil-crisis management, though in the latter category the Community essentially serves as a backup mechanism for the IEA (extending that mechanism to France).

But there are a number of disappointments. The Communities still spend a small fraction of their total budget (about 3 percent) on energy, whereas much more is spent, for example, on agriculture (71 percent). For several years the Commission has sent proposals to the Council on measures to aid the construction of coal-burning power stations, intra-EC coal trade, and the financing of coal stocks. None of these measures has been approved. In nuclear energy domestic opposition to nuclear programs has delayed progress; but much more could be done at the Community level. The critical area of energy-pricing policy remains to be addressed. On the investment side, the Commission needs to improve its mode of operation in encouraging the commercialization of energy sources and techniques, and to invest more in RD&D.

As a diplomatic instrument, the Communities' potential for North-South cooperation is strong. The EC have a favorable image based on their positive record of cooperation with associated countries under the Lomé Convention. Foundation has been laid for a Euro-Arab dialogue. The Communities are also useful as a framework for consultation with France before issues are discussed in the IEA, and with the smaller countries before summit meetings of the Western Seven.

In sum, the Communities are a valuable forum, but they could be much more active in energy policy if the member states would so allow. Unfortunately, given the sensitivity of the energy sector to national sovereignty, a common energy policy is inhibited by many political constraints.

Notes

1. An earlier version of this chapter was presented at the Conference of Europeanists sponsored by the Council of European Studies, Washington, D.C., October 23-24, 1980. The quotation comes from an official of the EC Commission interviewed by the author, January 1981.

2. See, for example, John Paxton, *The Developing Common Market* (Boulder, Colorado: Westview Press, 1976). For an early analysis of Euratom's performance, see Lawrence Scheinman, "Euratom: Nuclear Integration in Europe," *International Conciliation* (New York: Carnegie Endowment for International Peace, 1967).

3. "Energy Policy in the European Community: Perspectives and Achievements," Communication from the Commission to the Council, (Brussels: July 10, 1980), COM (80) 397 final.

4. The information in the following paragraphs is based on "Energy Policy in the European Community: Perspectives and Achievements," Communication from the Commission to the Council, COM (80) 397 final, (Brussels: July 10, 1980). See also "The European Community and the Energy Problem," *European Documentation,* 2 (1980).

5. Council resolution of June 9, 1980, in *Official Journal of the European Communities,* C 149/1-2 (June 18, 1980).

6. Council resolution of June 9, 1980 concerning new lines of action by the Community in the field of energy saving, in *Official Journal,* 149/3-5 (June 18, 1980).

7. "Member States' Energy Savings Programmes: Situation May 1980," Report of the Commission, COM (80) 899 final, (Brussels: January 26, 1981).

8. See "Energy Policy in the European Community: Perspectives and Achievements," COM (80) 397 Final (Brussels: July 10, 1980), pp. 13-16.

9. The foregoing is based on an interview with EC officials, Brussels, January 1981. Interestingly, when asked if the R&D budget was sufficient, the official responded that it was "almost enough. Double the present amount would be hard to manage efficiently."

10. Information in the following paragraphs is derived from "Energy Policy in the European Community: Perspectives and Achievements," COM (80) final, (Brussels: July 10, 1980), pp. 18-21, 5-6.

11. The relevant Council Directives are: 68/414/EEC, O.J. L 308, 23.12.68; 72/425/EEC, O.J. L 291, 28.12.72; 73/238/EEC, O.J. L 228, 16.8.73; 77/706/EEC, O.J. L 292, 16.11.77; 77/106/EEC, O.J. L 61, 5.3.77.

12. Meeting of the Energy Council, November 27, 1980, Council of the European Communities, Press Release 11742/80.

13. "The European Spot Markets for Oil Products," Communication from the Commission to the Council, November 12, 1980, COM (80) 707 final.

14. This example, and some of the preceding discussion, are taken from my earlier article, "Energy Policy in the Community," *The Annals of the American Academy of Political and Social Science,* 440 (November 1978), p. 120.

15. For a discussion of the history of the Euro-Arab dialogue, see Hanns Maull, *Europe and World Energy* (London: Butterworth, 1980), chapter 12.

16. Ulrich Everling, "Possibilities and Limits of European Integration," *Journal of Common Market Studies,* 28:3 (March 1980), p. 224. Reprinted with permission.

12 U.S. Energy Policy

Paul S. Basile

At the beginning of the 1980s, many Europeans and Japanese, among others, looked elsewhere than to the United States for world leadership. Like a still-groggy Gulliver, the United States following the second oil crisis seemed to many in the world like a once-powerful giant rendered helpless and immobile by the Lilliputians of OPEC, other Third World countries, and various of its own intransigent citizens. In part, the decline and fall of the United States was unavoidable. The world had grown both more complex and more mature: simple hierarchical global power structures (with the United States at the top) were no longer possible. And in part, the United States was contributor to its own undoing; a festering political sore, remnant of (among other things) the cumulative humiliations of Vietnam and Watergate, was allowed to challenge the efficacy of the world's leading democracy.

Nothing seems to have contributed more to this "crisis of confidence"—as President Jimmy Carter, in 1979 so poignantly, and perhaps impolitically, put it—than the oil crises of 1973-1974 and, more dramatically, of 1979. The image of a great power standing idly by while a recognized ally (the Shah of Iran) falls to revolution, the price of oil doubles in less than twelve months, and American diplomats are taken hostage burns sharply in the minds of many Westerners. Why was the United States so helpless? they wonder. Why has the United States had no energy policy to prepare for this?

To a degree, the U.S. problem has been ignorance, or naivete. Only now are certain energy facts being slowly absorbed into American thinking and policy processes. The United States is a major importer of oil, yet several surveys in the late 1970s revealed that most Americans were unaware that much of their oil was imported. Fewer still understood how higher prices could reduce imports, or comprehended the economic, financial, and security implications of such large dependence on one of the world's most politically unstable regions (the Middle East). And almost no one seemed aware that the rapid depletion of cheaply accessible energy resources would inexorably push prices up to the real values of the new, expensive, replacement sources.

The sudden realization of these uncomfortable truths may seem to momentarily paralyze energy policy. Yet it is not completely accurate to say that the United States, following the second oil crisis, has no energy policy. In fact, the United States has, and has long had, *too many* energy policies.

U.S. energy policies at the end of the Carter administration can be summarized not as a single theme but as a short list of dominant directions:

Higher energy prices. Decontrol of crude-oil and natural-gas prices in the United States is an established fact—probably dwarfing the demand and supply impacts of all other policies.

Government support of synthetic fuels. Financial support, both for research and development and for demonstration of domestic sources of unconventional oil such as shales and synthetic liquids from coal, seems assured.

Continued government regulation. Though the government seems to be giving the energy industry more free rein in several ways, too many other interests need attention. Utilities continue to be heavily regulated; various conservation initiatives are regulatory in design.

Bits and pieces of conservation. A variety of tax breaks, tax credits, loans, and grants are available for conservation measures.

Possibly, an energy-export-oriented future.

Reduced oil imports. Actually a goal or objective rather than a policy, the reduction of oil imports is worthy of mention here, as it provides a unifying theme for policy directions. Most proposed policies seem to have oil-import reduction as at least their stated intent.

Cooperation with allies. Having led in the establishment of the International Energy Agency, the United States will maintain its commitment to dialogue, information exchange, and oil sharing among IEA members in case of shortages.

In this chapter U.S. energy policies will be neither defended nor attacked. Indeed, they will not even be comprehensively described. Rather, an attempt is made here to review briefly the U.S. energy situation and energy history, with some irreverant observations on the political system that contributed so centrally to that history, followed by some selected descriptions of energy laws and enactments. These sections serve as the basis for the projections of future policies in the chapter's last section, "The Policy Drift."[1]

Surveying the U.S. Energy Situation

The United States consumes a lot of energy, as everyone knows. Americans use more natural gas, for example, than the Germans and Japanese together consume of *all* types of energy. Energy use per capita in the United States is exceeded only by that of Canada, Luxembourg, and two or three small, energy-intensive economies. The average Swede or German uses one-half to two-thirds as much energy as the average American. The United States consumes more oil than all of the energy used by all of the developing countries. The United States imports nearly as much oil as is produced in all of the oil-importing developing countries. The United States also produces a lot of energy—nearly 80 percent of its total needs. America produces more oil than is used by all developing countries.

Still, the United States imports oil—about 6.5 to 7 million barrels per day, or nearly 40 percent of its oil needs. And import dependence had been growing each year until about 1978.

Table 12-1 presents many of the major statistics of energy supply and use in the United States in recent years. The general situation of high U.S. energy use, increasing oil imports, and declining production deserves a closer look. These trends changed after the second oil crisis.

In 1978, oil imports to the United States were nearly 50 percent of oil needs. In 1979, the import figures were closer to 40 percent. Also, for 1978 to 1980, total oil consumption in the United States dropped by roughly 5 to 10 percent each year. To some extent, the fall-off in oil consumption can be attributed to the 1979-1980 recession, but much more of the reason is the direct effect of higher energy prices. Global energy prices doubled in 1979, at the same time that the United States was embarking on a decontrol of energy prices at home. These rapid increases in price, coupled with the expectation of higher prices in the future, induced a massive shift to small (and generally foreign) cars, to the chagrin of the "big three" United States automobile makers; a reduction in the amount of travel by automobile; and a proliferation of efficiency improvements in homes such as shifts away from oil to other fuels and so on. Many of these changes are permanent and are just the beginning of additional measures to be taken by Americans now acutely aware of the changed economics of saving energy when oil prices are high.

These events are having an impact in negotiations at the International Energy Agency (IEA), where America's alleged profligate oil use was a favorite object of European and Japanese criticism. Now Americans are cutting down on oil use faster than most other IEA members—including Canada. Future oil-import targets, as established in IEA deliberations, are understandably affected by recent United States success in cutting back on oil consumption.

Even though some of the reduction in energy use in the United States during 1979-1980 must be attributed to the economic recession, future use of oil will probably continue its downward trend, even after economic recovery, for the following reasons. First, demographic trends suggest slower increases in the labor force, in numbers of persons of driving age,

Table 12-1
U.S. Energy Statistics, 1973-1980

	Year				
Type of Energy	*1973*	*1975*	*1977*	*1979*	*1980*[a]
Oil *(million barrels per day)*					
Consumption					
Gasoline	6.7	6.7	7.2	7.0	6.5
Distillate	3.1	2.9	3.4	3.3	2.8
Residual	2.8	2.5	3.1	2.7	2.5
Other	4.7	4.2	4.7	5.3	5.0
Total oil consumption	17.3	16.3	18.4	18.3	16.8
Production					
Domestic crude oil	9.4	8.6	8.7	8.5	8.6
Domestic natural gas liquids	1.7	1.6	1.6	1.7	1.6
Imported crude oil	3.2	4.1	6.6	6.2	5.2
Imported products	3.0	2.0	2.2	1.6	1.6
Total oil production	17.3	16.3	19.1	18.3	17.0
Natural Gas *(trillion cubic feet)*					
Consumption					
Industry	10.4	8.6	8.6	7.8	8.3
Residential commercial	7.3	7.6	7.5	7.8	7.5
Utilities	3.4	3.0	3.1	3.4	3.5
Total natural gas consumption	22.0	19.5	19.5	19.0	19.3
Production					
Domestic	21.7	19.2	19.2	18.7	18.5
Imports	1.0	1.0	1.0	1.2	0.9
Total natural gas production	22.7	20.2	20.2	19.9	19.4
Coal *(million short tons)*					
Consumption					
Industry	68	64	62	58	62
Utilities	389	406	477	526	560

Coke and other	105	92	86	85	75
Total coal consumption	562	562	625	670	697
Production					
Domestic	599	655	697	760	810
Exports	53	66	54	63	85
Total coal production	652	731	751	823	
Electricity *(trillion kWh)*					
Consumption (sales)					
Industry	.7	.7	.8	.8	.8
Residential/commercial	1.0	1.0	1.1	1.2	1.2
Total electricity consumption	1.7	1.7	1.9	2.0	2.0
Supply (busbar)					
Oil-fired electricity	.31	.29	.36	.31	.25
Gas-fired electricity	.34	.30	.31	.34	.35
Coal-fired electricity	.85	.85	.99	1.1	1.2
Nuclear electricity	0.8	.17	.25	.26	.25
Other	.27	.30	.22	.28	.29
Total electricity supply	1.9	1.9	2.1	2.3	2.3
Nuclear					
As percentage of electricity generated	4.5	9.0	11.8	11.4	10.7
Capacity *(million kWh)*	—	—	43.0	49.0	53
Number of plants					
In operation	—	—	67	71	74
Under construction or on order (at year end)	—	—	132	116	102
Total number plants	—	—	221	190	176

Source: Adapted from U.S., Department of Energy, *Monthly Energy Review*, 1980 and 1981.
[a]Estimated.

and in new households. Second, we will probably continue to see slower rates of growth in the gross national product in the United States and in most of the Western world, and this will dampen domestic demand for energy. Third, there are shifts under way to substitute less energy-intensive processes and new technologies for the old energy-intensive ones. And price increases will continue to provide incentives for households and businesses to change their activities—to use car pools or mass transportation, to install insulation in buildings, and so on.

Other prognosticators may not agree that U.S. oil consumption will never again be as high as it was in 1979. But estimates of future use of energy in the United States have declined in the last several years—even

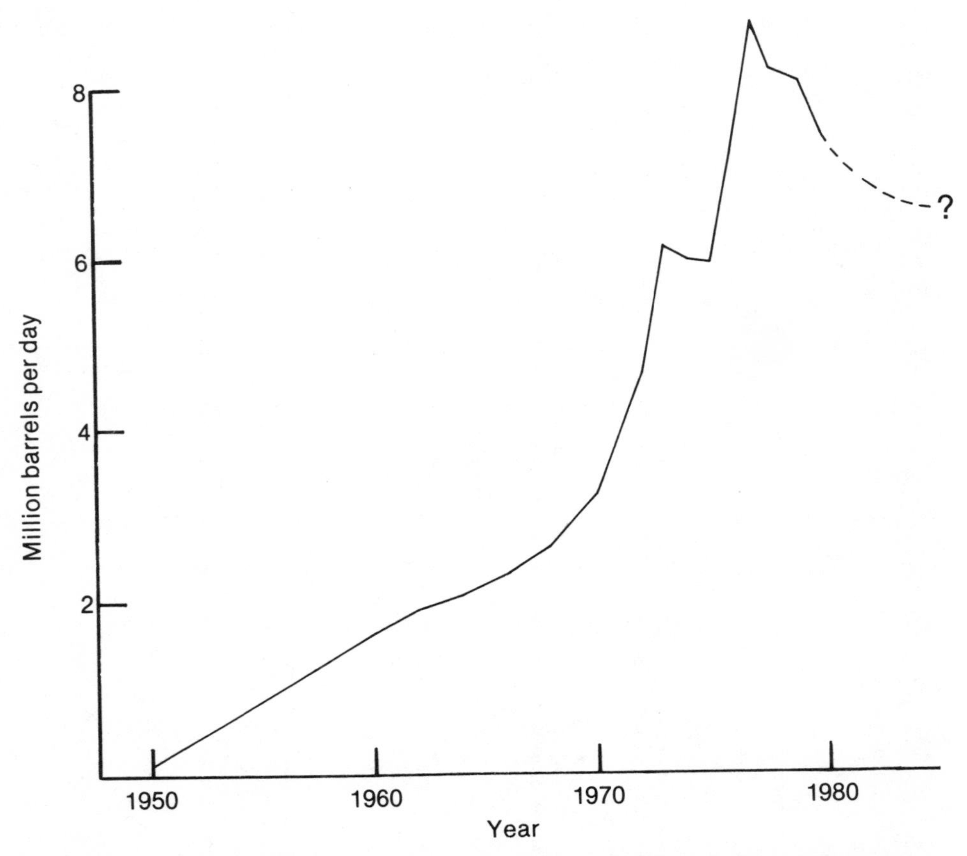

Source: U.S., Department of Energy, Energy Information Administration, *Annual Report to Congress* Vol. 2, Washington, D.C.: 1979.

Figure 12-1. U.S. Oil Imports

faster perhaps than U.S. prestige in the world. In 1972, the conventional estimate of United States' demand for energy was about 160 quadrillion Btus, or about 70 million barrels per day of oil equivalent. At that time, Amory Lovins was seen as heretical for forecasting only 125 quadrillion Btus of energy demand in 2000. By 1975, however, the forecasters were

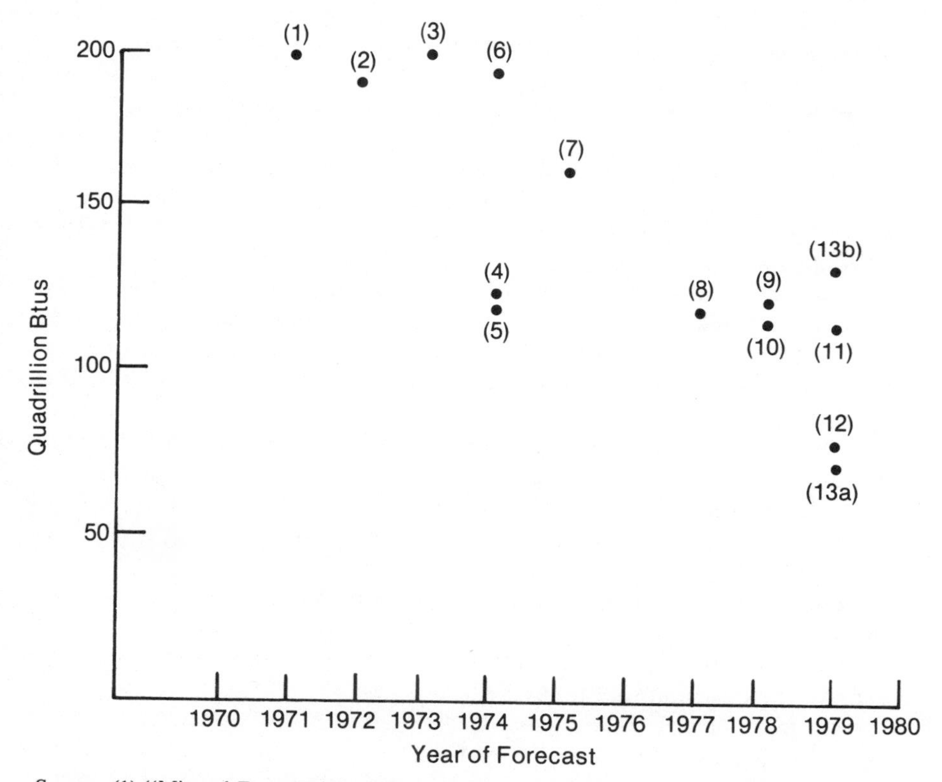

Source: (1) "Mineral Facts and Problems," *Bureau of Mines Bulletin*, 650, 1971; (2) *United States Energy through the Year 2000*, Dupree and West, U.S. Department of Interior, 1972; (3) *U.S. Energy Outlook: Energy Demand*, Energy Demand Task Group, National Petroleum Council, 1973; (4) *A Time to Choose: America's Energy Future*, Ford Foundation Energy Policy Project, 1974; (5) Council on Environmental Quality, 1974; (6) U.S., Atomic Energy Commission, *Nuclear Power Growth, 1974-2000*, 1974; (7) Energy Research and Development Administration, *A National Plan for Energy RD&D*, Vol. 1, 1975; (8) *Workshop on Alternative Energy Strategies* (McGraw-Hill, 1977) (Case C-2); (9) U.S., Department of Energy, Energy Information Administration, typical figure used in analyses, 1978; (10) U.S., Department of Energy, "Market-Oriented Program Planning Study," 1978; (11) *Energy: The Next Twenty Years*, Ford Foundation/Resources for the Future, Ballinger, 1979; (12) Alternative Energy Futures Study Team, Stanford University, 1979; (13) CONAES, *Alternative Energy Demand Futures to 2010*, report of the Demand and Conservation Panel, 1979. (Figure is for approximate range.)

Figure 12-2. Estimates of Energy Use in the United States in the Year 2000

pulling down their future estimates in the wake of OPEC price hikes, and the conservation movement it had induced. By then, 125 quadrillion Btus was a reasonable figure. But by then, of course, Amory Lovins had a forecast of 90 quadrillion Btus. Today, in the early 1980s, the range of estimates is wider, not narrower. Conventional wisdom might set the year 2000 energy demand at about 125 quadrillion Btus. But it seems that 100 quadrillion is a high figure. The conservationists are optimistic and estimates as low as 33 quadrillion Btus by the year 2000 can be found.

Oil use can continue to drop if it is recognized that shifting uses *away* from oil to other energy sources—even if the total energy used is greater after such a shift—may be more advantageous than single-minded conservation of any one kind of energy. On the production side, there seems little likelihood of increasing or even holding constant oil production in the United States in the 1980s. Dwindling reserves have already created a reserve-to-production ratio of about 7.5, making rapid production increases unimaginable. In the longer run, of course, the United States has abundant resources of shale oil and perhaps of other difficult and expensive oil sources that might be tapped. However, these costly sources have long lead times and entail a variety of institutional, environmental, social, and other obstacles.

The United States' production of natural gas has many of the same characteristics as its oil production. Yet with the gradual phasing out of controls on wellhead prices of gas, many believe that abundant new gas sources will be found and that it may be possible to hold gas production at its 1980 level. That would be an optimistic forecast.

Of course the United States has abundant coal resources, but coal is not the most preferred fuel. Environmental objections, pollution problems, and other regulatory restrictions must be dealt with if coal production is to achieve the potential that many believe it has.

No new nuclear plants have been ordered in the United States since 1978. In the late 1970s, thirty-nine of the original orders for nuclear plants that had been placed in the early 1970s were cancelled. Yet nuclear power provided in 1980 some 13 percent of electricity supply in the United States, and may provide 20 percent by 1985, based on plants in construction in 1980. Unresolved technical, political, and institutional problems continue to plague the nuclear power industry in the United States. Given the abundance of alternatives for producing electricity and the apparent rapid decline in growth of electricity demand, it is not likely that there will be any significant increase in new orders of nuclear plants in the first years of the 1980s.

Coupled with the public's response to the high prices of oil and the improved efficiency of energy use are shifts away from oil to other fuels—particularly, where it makes sense, to renewable sources such as wood or solar power. Wood is a major energy source in the United States today and its use for home heating is increasing very rapidly. Solar technologies are proliferating at a great rate and their penetration is being encouraged by extremely favorable tax laws.

Although this largely statistical picture of the United States energy situation sets a background, it does not capture the complex evolutionary trends in America that are so essential to understanding its energy policies. The next section reflects on the role of oil companies in America, the development of government policies, and the nature of the U.S. political system.

Seeds of National Confusion

The Oil Companies

One fact more than all others sets the United States apart from other industrialized energy-consuming countries. The United States is, and has been, a substantial oil producer. Indeed, wealth from oil transformed America, as such wealth is transforming the developing oil-producing countries today.

The oil industry is rooted in Texas, Oklahoma, and Pennsylvania. The barrel, today the dominant international measure of oil quantity, originated as the wooden barrel used for oil in Pennsylvania in the 1860s. It is only very recently that control of the oil business has been uprooted from its tightly-held U.S. oil-company base and transplanted to the Middle East.

From their beginnings, the oil companies have had a love-hate relationship with the U.S. government (and public). The trust-busting of eighty years ago did not seem to diffuse power very much; the subdivided companies cooperated more than they competed. And although the companies earlier built up worldwide networks of distributors and agents, nearly all their production came from the United States. It was not until World War I that the companies began to look seriously outside the United States for supplies.

The historical role of the oil companies is important in understanding U.S. energy policy after the second oil crisis for at least three reasons. First, it helps explain why the United States developed such oil-dependency: the U.S.-based companies succeeded in holding a substantial degree of influence, if not control, over market price and quantity. Second, the government generally dealt kindly with the companies, being reluctant to tax a domestic resource used for domestic consumption. (Notice that OPEC and other oil-producing countries today that charge $30 and more per barrel for exported oil often charge much less—$1 or $2 per barrel—for oil used at home.) Third, the American people are well aware (probably overly aware) of oil-company influence, they distrust the companies because of it, and they are unwilling to support actions that might be wise policy but at the same time are beneficial to the "robber-baron" companies. The public is also unwilling to accept the recent fact of transfer of control away from the companies: they believe there must be some conspiracy between OPEC and the companies.

Polls in 1980 showed that three-quarters of the American public agree that there is an energy crisis—and that one-half of them blame it on the oil companies. For example, during 1979 one used to frequently hear statements such as "just wait until gasoline prices top $1 per gallon—*then* you will see the oil companies flood the market with gasoline!"

The history of attempted control of the oil business, both globally and in the United States, only began to be revealed during a series of U.S. Senate hearings on multinational corporations in 1952.[2] Control of oil in the United States surfaced as an issue as early as 1929, when the American Petroleum Institute approved a plan to restrain production of crude oil in the United States to the levels of 1928. And the Texas Railroad Commission (a state government agency) found ways to literally dictate the oil business in their state. Venezuelans, in fact, were foresighted enough in the 1930s to hire a man from the Texas Railroad Commission to advise them.

Globally, the attempt at oil control by the major oil companies was governed by the philosophy of As Is—that is, keeping shares of oil properties and future production at then-current proportions, thus excluding new competitors. This was, however, never fully achieved. Nevertheless, the appearance of this attempt in the United States—with its anti-trust history and built-in antagonism between government and private industry—helped to ensure the later public difficulties of the oil companies, so much a factor in the energy-policy process today.

This whole fabric of relationships between the oil companies, the United States government, and the public is one that is not well-understood by Europeans, because they have never been oil-producing countries (the United Kingdom is the recent exception).

Official U.S. government policy toward the oil companies has always been fragmented. While the Justice Department continued for nearly a century in their popular trust-busting activities, another part of the U.S. government was quietly delegating foreign-policy initiatives and responsibilities to the oil companies. For years this seemed to work all right. The public foreign-policy position of the United States in the Middle East was that of support for Israel. The quiet, company-directed foreign-policy position was of support for oil-rich, West-leaning, Soviet-fearing Saudi Arabia. Only with the new-found wealth and power of the Arab countries of the Middle East has this tentative and awkward web of responsibilities begun to unravel.

The year 1953 was a particularly active one on both sides of this confrontation. One report from the U.S. Department of State to the National Security Council, a report backed by the Defense and Interior departments, boldly supported the case for strong government support of the major oil companies. It said that the companies "play a vital role in supplying one of the free world's most essential commodities" and added, "American oil operators are for all practical purposes instruments of our foreign policy towards these countries."[3]

At the same time, the Attorney General presented his case to the National Security Council, arguing that the companies, far from being crucial to national security, were profoundly damaging to it. The report stated: "It is imperative that petroleum resources be freed from monopoly control by the few and be restored to free, competitive, private enterprise."[4]

Here were the classic horns of a dilemma. To take stronger anti-trust action against the companies would be to invite a takeover of control by the oil-producing countries; an event which we now know was inevitable. Conversely, not to take action against the companies would be to admit a failure of democratic institutions and to condone the existence of a multinational cartel. For a long time, the companies quietly dominated energy policy in the United States through state-by-state production control, encouragement of consumption at a steady pace, and the holding of prices at what then were relatively high levels in order not to be undercut in price by the much cheaper production costs of Middle Eastern oil. Even in the 1960s the U.S. oil companies were losing control of the market because of lower-priced foreign oil.

At about the same time, early in the 1970s, the U.S. oil-supply situation became problematic. U.S. oil production could no longer increase to meet demand, imports of oil to the United States were suddenly much more costly with OPEC's quadrupling of prices in 1973-1974, and vulnerability to foreign suppliers was front-page news with the Arab oil-embargo. The quietly inserted oil-price controls in the United States (originally put in place as ceiling prices to protect the oil companies' domestic United States production) came under attack from oil companies as holding the price artificially low and discouraging incentives for further production.

The energy analyses of the mid-1970s or 1980 seem at best confusing. But the confusion has explainable and even justifiable historical contexts, as I have attempted to show. Energy policy in the United States has had to cope with the situation at hand. Complicating the situation at hand is a political system as unlike those of Europe and other industrialized countries as is the oil supply picture in each case. The profound differences are not well understood.

The Political System

To say that U.S. politics, oil companies, and energy policy are all interrelated is to say nothing new; but long-standing public and political attitudes weigh more heavily than one might initially think. One large reason for this is the role and structure of an increasingly assertive, public, and political institution: the Congress.

Congress mirrors the public. That is its job, to an extent. And the U.S. public is extraordinarily pluralistic—in race, creed, religion, culture, social stratum, and so on. The well-known American individualism simply accentuates the pluralism by reinforcing the desirability of differences. But carried too far, this diversity can tend to paralyze action. It can subvert the general welfare to individual self-interest. It can move decision making toward coalitions of individual self-interests and away from collective reasoning for national interests. This painfully but accurately describes the U.S. Congress today. And Congress is where U.S. energy policies are made (or not made).

U.S. political parties are not like the parties of Europe's parliaments. No party is in power in the parliamentary sense of "forming a government." The tradition-bound American separation of powers, of checks and balances between the executive and legislative branches, means in practice that no one branch really governs. Neither the administration nor Congress is accountable for what goes on: each blames difficulties on the intransigence and narrow-mindedness of the other. President Wilson wisely observed that "power and strict accountability for its use are the essential constituents of good government." In the United States, accountability is hopelessly fractured.

Government officials appointed by the President—the Secretary of the Department of Energy, for example—*do not* simultaneously hold seats in Congress (as they virtually always do in parliamentary systems) and almost never have any party role. This typically exacerbates the enmity between the executive branch and the legislature, and further impedes policymaking.

The divisions among members of the executive branch seem also to be exacerbated by the system. James Schlessinger, when Secretary of the U.S. Department of Energy (DOE), enjoyed the confidence of the President, but apparently of no one else. His replacement by Charles Duncan seemed to many to signal that energy policy would be lodged more firmly in the White House—with Lloyd Cutler, Kitty Shirmer, and others—and less in the DOE. The wounds can be deep and hard to heal. President Ronald Reagan, at this writing, seems intent on further circumventing DOE.

Much has been said and written of Congress' "new assertiveness." Following the Nixon years of the "imperial presidency" and the collective impotence of Congress during the Presidentially-escalated nondeclared Vietnam war, Congress has sought to reestablish its primacy. It is taking new pride in its independence from the White House, and has greatly increased its own professional staff in the last few years. (Unlike a European Member of Parliament, who has little or no staff, a U.S. congressman will typically have a full-time staff of about fifteen people; and each of nearly two hundred standing committees and subcommittees in the House of Representatives, for example, may have a staff of ten to sixty or more.)

But Congress, true to the national character, is itself a highly pluralistic, diverse institution with a bewildering array of individualistic and often conflicting forces. As President Carter's own legal counsel has observed:

> The former ability of the President to sit down with 10 or 15 leaders in each House and to agree on a program which those leaders could carry through Congress has virtually disappeared. The committee chairmen and the leaders no longer have the instruments of power that once enabled them to lead. . . . All this means that there are no longer a few leaders with power who can collaborate with the President, power further diffused by the growth of legislative staffs. . . . In the past five years the Senate alone has hired 700 additional staff members, an average of seven per member.[5]

The style is confrontational: opposing sides (usually more than two) stake out the extremes. Then, they stage a long battle—a war of attrition. There seems to be no built-in process of accommodation or cooperation.

In addition, Congress deals with a great amount of detail. This further tends to blur and complicate the search for broad policy directions. For example, after months of separate debates in the House of Representatives and in the Senate on the Synthetic Fuels Corporation and on the Energy Mobilization Board proposals of the administration, the joint House-Senate Conference struggled for six months with the highly detailed and cumbersome procedures before ultimately reaching agreement—and then the agreement on the Mobilization Board was rejected by the House.

These rather discouraging descriptions of the political process in the United States legislature are not intended to cast blame for all energy-policy failures on the Congress, nor to paint the legislature totally black. Indeed, there are a number of highly intelligent, extraordinarily hard-working and well-meaning individuals who honestly and conscientiously seek to serve the national interest. But the *structure* of the American political system—and specifically of the Congress and its relationship with the executive branch—is no longer able, in today's complex world, to cope with the important policy issues that we face.

If the U.S. political system cannot be designed for stronger political parties, or more directly accountable government, perhaps the dire seriousness of the energy and other issues we face will mobilize a coalition of large segments of the American public and hence of the Congress to take decisive, wise, and collective action.

The New Dimension

Reinforcing the unfortunate trends toward diffusion of accountability and lack of workable processes to collectively identify the common good is a new dimension in American policy: the intervenors. Intervenors are *single-issue interest groups*—environmental, pro-oil-company, pro-utilities, anti-nuclear, pro-solar, and on and on. These are the groups which can, with tiny and sometimes uninvolved minorities, through the complex and overgrown United States legal and regulatory systems, totally tie up energy projects, energy bills, and energy policies.

Protecting the interests of potentially disadvantaged individuals and groups is necessary and desirable; when it impedes all activity the devices for such protection must be reassessed.

There are underlying reasons for the growth of influence of intervenors. Energy projects that have national benefits often carry regional costs. Thus, oil production from the shale rock of northwestern Colorado makes use of a domestic resource, creates jobs, reduces the economic drain of importing oil, improves the balance of payments and the strength of the dollar, and so on—all national benefits. At the same time, the oil-shale business creates problems of the environment, water supply, social upheaval, and other regional problems. More and more, a limited regional population is asked to bear the burden of costs for national benefits: communities close to nuclear power plants, Alaskans near oil-production or pipeline facilities, coastal residents near shipping lanes.

Analogously, certain groups are disadvantaged and others benefited for net national gain. Indeed, the real policy choices are not clear options, but trade-offs: higher oil prices and reduced import dependence versus unavoidably higher energy bills for the poor or fixed-income elderly; expanded energy-production incentives versus environmental protection; military support for Saudi Arabia to secure strategic supplies versus military support for Israel. Always, it seems, some groups are hurt when others are helped. National energy policies must face the realities of regions and groups fated to carry the costs. Increasingly and understandably, those burdened with costs object and intervene.

Interregional problems in the United States are not seen by those abroad seeking a unified U.S. energy policy. New England, for example, is an energy-consuming region—no indigenous resources, high population, and a struggling economy that gains no apparent benefit from high energy prices. Texas, on the other hand, is an energy-producing region—plenty of resources, low population, and an economy benefiting from high energy prices and controlled production. Forging a national energy policy with such interregional forces might be akin to building a consensus energy policy between France and Saudi Arabia. It is not easy. Yet the trade-offs must be made. The alternative is stalemate—delay and inaction. The nation suffers, and ultimately so will each region and group.

These are deep issues in a democratic society, without ready answers. The hope is that solutions will be found by reason and fairness, not by the force of oil shortages and national-security pressures, or by a third oil crisis.

United States Energy Legislation: A Tale of Two Crises

Energy Legislation prior to the First Oil Crisis

Prior to the first oil crisis in 1973, it was generally believed that energy was cheap and abundant. The legislative energy agenda in the United States during this long and relatively stable era was to ensure that energy was not too cheap (oil prices during this period were typically below $3 per barrel). If producers were allowed to produce all that they could, they would overproduce (resulting in the feared oil glut) and drive prices too low. An example of the price-floor legislation enacted during this period is the Connally Hot Oil Act which prohibited interstate sale of oil produced in excess of state allowances. Thus states were able to control the amount of oil sold and produced with state allowances. Even as late as 1959, when it appeared that oil produced in countries outside the United States could be sold for less than oil produced within the United States, the Mandatory Oil Import Program was enacted to discourage importation of cheap foreign oil that competed with U.S. sources. These bills and others are listed chronologically and briefly characterized in table 12-2.

In addition, there was in this period a growing federal role in energy matters. The federal government was to regulate interstate commerce and prevent overpricing to customers. There was an increased reliance on technology for future energy supply (the Atomic Energy Act of 1954, the Atoms for Peace program, and promised power "too cheap to meter"), and little, if any, erosion of the oil companies' favored position. The famous—or infamous—depletion allowance recognized the companies' contention that they dealt in a wasting asset.

Those sharing a widespread belief in the abundance of cheap and accessible energy—undiminished by the fact of already high oil imports to the United States in the early 1970s—were due for a rude awakening.

The First Oil Crisis, 1973-1974

In 1973, the United States had just experienced a period of what was then considered rapid inflation. President Richard Nixon had recently imposed price and wage controls on the economy in an effort to slow the inflation. Into this foray came the first oil crisis; the oil-producing countries in a span of six months raised the price for imported crude oil by over 300 percent. Arab producers briefly compounded the problem (and made the high prices stick) by embargoing the United States and other supporters of Israel in its war effort. The increased import prices created problems in the United States significantly different from those of its trading partners in Europe

Table 12-2
Chronology of Major U.S. Energy Legislation

Year	*Act*	*Description*
1889	Rules of Capture	Common Law: granted landowners the right to recover any and all oil from wells drilled through the surface of their properties.
1920	Federal Power Act	Established Federal Power Commission which regulated electric utility companies engaged in interstate commerce.
1926	Depletion Allowance	Oil and gas producers were allowed to deduct 27.5 percent of gross income from oil production.
1935	Connally Hot Oil Act	Prohibited the interstate sale of oil produced in excess of state allowances.
1935	Public Utility Holding Company Act	Provided for control and regulation of public-utility holding companies.
1938	Natural Gas Act	Directed Federal Power Commission to regulate interstate gas companies.
1954	Atomic Energy Act of 1954	Established Atomic Energy Commission.
1954	*Phillips Petroleum* vs. *Wisconsin*	Supreme Court interpreted Natural Gas Act to give the federal government power to control wellhead prices for natural gas.
1959	Mandatory Crude Oil Import Program	Limited the amount of oil imported.
1968	Natural Gas Pipeline Safety Act	Established safety standards for pipelines.
1969	Tax Reform Act	Reduced the percentage of oil depletion to 22 percent.
1973	Emergency Petroleum Allocation Act	Established domestic crude-oil and oil-product price-and-allocation controls.
1973	Trans-Alaska Oil Pipeline Act	Authorized and supported construction of the Alaskan oil pipeline.
1974	Energy Supply and Environmental Coordination Act	Required reports with respect to energy supplies, provided for temporary suspension of air pollution requirements, and for coal conversion.
1974	Federal Energy Administration Act	Established the Federal Energy Administration.
1974	Energy Reorganization Act	Established the Energy Research and Development Administration.
1974	Federal Nonnuclear Energy Research and Development Act	Provided funds for energy research and development.
1975	Tax Reduction Act	Eliminated percentage of oil depletion allowance for majors and decreased it for others.

1975	Energy Policy and Conservation Act	Extended oil price controls, established a Strategic Petroleum Reserve, required conservation plans, set automobile fuel-efficiency standards, and established conservation assistance.
1976	Alaska Natural Gas Transportation Act	Set the groundwork for a presidential decision on the Alaskan natural gas pipeline.
1976	Energy Conservation and Production Act	Provided for production incentives, energy-conservation standards for new buildings, and energy-conservation assistance.
1977	Department of Energy Organization Act	Established the U.S. Department of Energy.
1978	Petroleum Marketing Practices Act	Protected retail petroleum-dealer rights.
1978	Natural Gas Policy Act	Ended bifuricated pricing, restructured the pricing system, and gradually decontrolled natural-gas prices.
1978	Power Plant and Industrial Fuel Use Act	Required gradual conversion of oil and gas utility and industry users to coal and alternative fuels.
1978	Public Utilities Regulatory Policies Act	Set voluntary guidelines for utility pricing structures.
1978	Uranium Mill Tailings Control Act	Regulated uranium-mining by-products.
1978	National Energy Conservation Policy Act	Provided grants, loan insurance, and consumer information for energy conservation.
1978	Energy Tax Act	Imposed a small tax on cars with very low fuel efficiency and established tax credits for energy conservation and alternative energy sources.
1979	Emergency Energy Conservation Act	Authorized the President to develop a gasoline-rationing plan and set conservation targets for states.
1979	Phased Decontrol of Domestic Oil Price Controls	By executive order, the President phased out controls on domestic crude oil.
1979	Low-Income Energy Assistance	Low-income households were given block grants and subsidies.
1980	Windfall Profits Tax	Levied excise taxes on various categories of domestic production to recover some revenue from phased decontrol and higher world oil prices.
1980	The Energy Security Act	Established the Synthetic Fuel Corporation (at a cost of $20 billion) and funded biomass, conservation, solar energy, geothermal energy, and renewable energy commercialization programs in lesser amounts.

and Japan. First, the United States still produced over half the petroleum it consumed. It was considered unfair that these American producers should benefit by receiving the artificially high prices charged by the OPEC countries. Second, although the United States' domestic oil market had (and still has) a multitude of competing firms—major oil companies, independent oil companies, and small oil companies—the major oil companies had the main access to the cheap foreign oil; giving them, it was thought, an undeserved competitive advantage. Third, consumers had come to treat inexpensive energy as a necessity.

The U.S. government reacted by controlling oil prices, as embodied in the Emergency Petroleum Allocation Act (EPAA) of 1973. This began a period of legislation designed to fully entrench the federal government in the energy business—controlling it wherever, in the highly heterogeneous American economy, it could find a way to do so.

The United States made some rather feeble attempts to produce its way out of the first oil crisis. Legislation was passed that attempted to increase conventional domestic oil production by offering incentives such as higher prices for certain categories of oil with, presumably, higher potential production. For example, new oil discovered after 1973 was priced higher than old oil discovered before 1973. Furthermore, investments in R&D for alternative-fuel projects were increased; seed money was given for a wide range of feasibility studies for various energy technologies. Large pipeline projects were also initiated in response to the first crisis. In 1974, the Trans-Alaska Pipeline Act was passed authorizing the construction of the Alaskan oil pipeline. In 1976, the Alaska Natural Gas Transportation Act set the groundwork for the later presidential choice of the pipeline as a method for bringing natural gas from Alaska to the lower states. (Construction of the Alaskan natural-gas pipeline has yet to begin.)

As price controls on United-States-produced oil continued, the lower prices brought the logical result: the United States increased the quantity of its oil imports. The low prices shielded consumers somewhat from the rise in international prices, and thus did little to encourage conservation; they did even less to encourage domestic production. Low prices in the name of consumer protection necessitated a spate of legislation to mandate certain energy-efficiency improvements and to subsidize consumer conservation.

Thus U.S. energy legislation during the period just after the first oil crisis was characterized by controls on energy prices and the allocation of energy supply. When the reaction to the controls proved to be high consumption and low production—or low activity in exploration and in seeking alternative sources of energy—the government stepped in to force or subsidize these latter operations. Automobile fuel-efficiency standards were put in place in 1975, and conservation programs were subsidized. However, most of these subsidies were relatively small. The real objective was to avoid direct price increases. Energy taxes, not surprisingly, were virtually unthinkable.

Moral Equivalents and the Second Oil Crisis

President Jimmy Carter came into office in 1977 hailing the solution of the "energy problem" as his number-one priority. In April 1977 he unveiled a series of measures designed to reduce the United States' dependence on unreliable foreign oil sources. He called for belt-tightenings and sacrifices, and soberly announced that the problem was serious enough to require the "moral equivalent of war."

The protracted debates following April 1977 confirm the contention that the U.S. political system is too fragmented to coherently come to grips with national needs. And a new and inexperienced President mishandled, by most accounts, the tactics of dealing and negotiating with a headstrong Congress—and, it seems, even dealing with the diverse interests within the executive branch.

The first step Carter took, not without rancor, was the creation of a single U.S. Department of Energy. In the following year, 1978, came the fruition of all the debate: seven major energy bills only remotely resembling the original Carter proposals.

Events in 1977 and 1978 had mobilized the legislators to take energy actions. The 1977 shortfall of natural gas signaled that the energy-supply situation was insecure. By the end of 1978, the fall of the Shah of Iran and the curtailment of Iranian oil production seemed to confirm this. The legislation of 1978 recognized the permanence of higher energy prices and began to grudgingly recognize the inability of government to control prices forever. The Natural Gas Policy Act of 1978 provided for a very gradual path to price decontrol of natural gas by 1985 (controls on oil prices remained). It was also recognized that electric utilities should convert from oil and gas to alternative fuels, and that new utility-pricing structures might be required to better reflect the scarcity of energy. The first breakthrough on the tax front was a very small tax on large fuel-inefficient cars. A modest proposed gasoline tax was vigorously rejected.

By 1979, the events that together signaled the second oil crisis made it quite clear that oil prices were going to rise, and they did. They also brought about the realization that the United States would continue to be vulnerable to import disruption and the associated quick price jumps. Thus, the Emergency Energy Conservation Act of 1979 was passed, which gave the President authority for gasoline rationing, and to set state conservation targets.

Consistent with the gradual decontrol of natural gas, the President in 1979, by Executive Order (and, one must say, given the peculiarities of the United States's oil situation, with considerable political courage), gradually decontrolled the price of domestic crude oil. The phase-out of price controls would be complete by October 1981. The shorter phase-out for oil than for gas was as much dictated by the structure of the EPAA in 1973 as it was an indication that the necessity for the end of cheap energy was finally being more widely realized. Yet decontrol, coupled with rising international oil prices, would (and did) bring sudden large profits to the oil companies. So the Windfall Profits Tax of 1980 was offered by the President as a necessary partner to decontrol. The tax captured some of the revenues from the higher prices and passed them to the U.S. Treasury.

The second oil crisis also reinforced the U.S. preoccupation with the vulnerability issue. Widespread opposition to registration for the military draft served to heighten any and all attempts to break what many believed to be the stranglehold of Middle Eastern oil producers on the United States. Tremendous amounts of political energy were spent on urging that the Strategic Petroleum Reserve be filled, and quickly. More than a few U.S. energy-policy experts listed strategic storage as the number-one energy-policy imperative.

In 1980 there was a substantial commitment by the U.S. government to the production of synthetic fuels, in the Energy Security Act. Under this act, the government began to give large subsidies to various forms of alternative energy supplies. Synthetic-fuel projects were given an initial subsidy of $20 billion, with a possibility of further subsidies of up to $68 billion. Smaller subsidies were also given for residential conservation, biomass-energy research, geothermal energy, renewable resources, and solar energy programs. In addition, in another piece of legislation, funds were set aside for assistance to low-income households faced with high energy bills.

Thus, U.S. energy policy in 1980 included higher prices, but compensated returns of the revenues to the government through higher taxes on production; and it included increased subsidies for alternative fuels and conservation and compensation to low-income households.

At the end of the Carter administration the United States was still feeling its way around in energy policies. The recognition of the failure of price controls and other regulations was taking hold. The energy market of the future is widely seen to be one of permanently higher prices and vulnerable supplies. America's future energy legislation will likely be characterized by decontrol of prices and handling of equity problems by taxation of production and redistribution of money to consumers—by subsidizing alternative energy sources. Future legislation may even see higher-than-market prices, again with equity concerns being dealt with through taxes and rebates.

The Reagan Administration seems likely to put greater emphasis on market forces through higher prices, and reduced emphasis on redistribution of growing revenues; there is even talk of repealing or at least scaling back the windfall-profits tax. It also appears that President Reagan is cool toward government subsidies of synthetic fuels. But it seems unlikely that the program would be totally shelved.

The Policy Drift

The Proper Role of Government

The last several years have been a time of searching in the United States: searching for information, searching for understanding, searching for some lever to pull that could make some difference in the destructive energy events illustrated by the second oil crisis, over which the United States seemed to have little or no control.

Should the government have any control over international energy events? Regrettably, there has never been a clear vision in the United States of what the government role *ought to be* in energy affairs—nor, for that matter, in foreign affairs. Should the government encourage activities beneficial to the nation through subsidies (loans, grants, price controls, etc.), or through tax credits or other tax incentives, or through wholesale demolition of bureaucratic red tape (some of which, one must not forget, is perceived as a lifeline by important parts of society), or through some new, imaginative forms of regulation at the federal or state levels. One thing seems clear: regulation in theory operates in vastly different ways from regulation in practice.

True, the government cannot responsibly let its economy be buffeted about by forces outside of its borders; yet neither can the government dictate actions unilaterally to entities outside its borders. What is required is a sensitive appraisal of international events—and interregional pressures at home—and, within that context, some clear and pragmatic thinking on what can be done.

The difficult question of government's role in energy matters is one that must be honestly and comprehensively addressed by Congress. It must be Congress, because only in this way can the *public* be fully involved. And it is the public after all, in the highly decentralized United States economy, who will through their actions make or break any initiatives proposed by Congress. Government should intervene in energy matters only in order to provide certain signals, incentives and disincentives that the market by itself might not offer, or to short-circuit any possible abuses or excesses. It should seek to address *equity* considerations by moderating (through redistribution of energy-production revenues to low-income consumers, for example) the income-distributing effects of the unimpeded energy market which, at best, provides aggregate national benefits. Always, certain groups bear disproportionate costs while others reap disproportionate gains.

Future U.S. Policies: Goals and Priorities

Although energy-policy initiatives and the role of government may not yet be perfectly defined, a few national energy goals seem to be coming into focus. Goals can be and often are stated in many ways, usually with closely similar meanings. American national energy goals include:

Reducing energy consumption (in particular, oil consumption).

Using fuels more efficiently.

Importing less crude oil (that is, reducing vulnerability to disruptions in oil supply).

Among the incentives to try to reach these goals are the facts that petroleum fuels actually have a higher value than has been assigned to them in the recent past (this should be reflected in their more frugal and efficient use); and that high levels of oil imports carry a high cost in lost income and jobs, downward pressure on the dollar, and threats to national security and international prestige.

While the goals and underlying premises may be clear, the actions needed to reach the goals are far from fully defined. U.S. energy-policy analysts disagree on many things. The one thing nearly all seem to agree on, however, is that we must keep all our options open. Not so. In spite of a ponderous political system, inbred public hostility toward industry in energy issues, and aftereffects of the shock accompanying loss of world power and prestige to the Middle East, the United States must quickly make some tough trade-off decisions.

Americans keep striving for comprehensiveness. President Nixon's naive Project Independence and President Ford's National Energy Act were supposedly rallying cries against the encroaching Arab oil leverage. President Carter's "moral equivalent of war" met the same fate as the previous two comprehensive policy attempts—that is, rebuff by a resistant Congress after years of protracted debate. President Reagan seems determined to scrap most recent energy-policy initiatives for a return to a laissez faire free-market policy.

The United States faces not a technological problem, but a political and institutional one. American energy-policy analysts continue to grope for the comprehensive solution. But it is not to be found.

What *can* be done? Most of the actions that have been taken are politically least sensitive and least harmful. They are also the actions that make the least difference. We have a 55-mile-per-hour speed limit in the United States; we have tax breaks and benefits for renewable energy and popular technologies like gasohol production, and we have more money for energy R&D. At the same time, crude-oil price decontrol has passed—and may soon be speeded up.

The rewards of taking tough action are inevitably delayed; lead times for major new energy projects are long. The length of time between seedtime and harvest is typically longer than the average congressman's term of office. A right decision may carry present-day costs and future benefits to be reaped by someone else.

Yet there are things that can be done and must be done. Here are some important priorities for the United States in the 1980s and 1990s in energy policy:'

1. A new, more sensitive, and balanced foreign policy.
2. A necessary margin of safety must be maintained in strategic oil stockpiles; vulnerability to future supply disruptions must be reduced.
3. Transportation is the first-priority use of liquid fuels; effective oil-consumption policies must focus on the transportation sector.
4. Natural-gas supplies in the United States can be plentiful. The whole system of price and quantity regulation needs to be completely overhauled.
5. The abundant coal resources of the United States must be developed quickly. The high cost of pollution control can and will be borne by the buyers of coal, who will still find it more than competitive with oil at present and conceivable future prices. The United States must recognize the possibility and even responsibility for exporting coal in quantity.
6. Renewable resources, such as solar and biomass energy, can play a role but should hardly be given the excessive incentives they were given under the Carter administration. They are still relatively expensive sources of small amounts of energy, though their ultimate potential may well be very great, if technological breakthroughs decrease costs.
7. The cheapest barrel of energy is the barrel of energy saved. This is now clear. Studies of the ways in which incentives can be readjusted so that consumers may take advantage of the real-dollar savings of reducing energy use could help to overcome the institutional barriers that now seem to impede the wider practice of saving energy.

Some things are not proposed as policy priorities in the 1980s and 1990s. It is not necessary to offer large quantities of government money to the synthetic-fuels industry. This is an industry that can develop all by itself—and will likely take so long to do so that government money spent today will hardly ever see a return. (Of course, *because* it will take so long, government aid is felt by some to be necessary.) The only suitable government role here is to guarantee purchases of fuels.

It has already been implied that large tax breaks, loans, and other money for the development of renewable energy sources and gasohol are misplaced. Oil-import fees or import quotas do not seem primary, although the gasoline tax could be used effectively. The nuclear issue is moot; the nuclear industry seems to be on the decline in the United States. In any case, strict safeguards are certainly a government responsibility.

Of course, nuclear power is a serious alternative for many industrialized resource-poor countries. But in the United States the alternatives are so numerous and so tractable, generally, that it is hard to see any compelling justification for pushing nuclear power where it seems to be resisted.

The clear choices for the United States are to: generate electricity with coal; focus on conserving energy in the transportation sector (not on providing fuels for energy-inefficient vehicles); put the right incentives in the right places so that people will become more energy-efficient; and, if necessary, build well-insulated, passive new solar buildings (natural gas can meet any remaining energy needs in buildings). America should forget the excessive emphasis on the oil-import problem and shift to a new, more sensitive international foreign policy. The United States should not seek to exploit shale resources now, but expand the coal industry. The strictest reasonable environmental requirements should be imposed while cutting permit and appeal time for new coal projects. If all energy prices are at market levels, coal can and will be sold.

Elaboration on some of these points may be in order. With regard to domestic initiatives, the United States's energy policies would be extremely well served by some thoughtful transportation-system planning. Some recent (to date unpublished) studies of the subject have demonstrated, compellingly and convincingly, that government dollars spent to revitalize northeastern industries, to build newer, smaller, more efficient cars, and to build newer and more efficient public transportation systems (which have the added effect of greatly improving urban quality of life), will reap much more substantial and immediate benefits than the same government dollars spent to promote synthetic-fuels production. Encouragement of the entrepreneurs and inventors working on greatly improved battery design for electric vehicles might be investments well made.

On the conservation side, government policies to date have been generally in the right direction. A variety of loans, tax credits, and other incentives are probably working in a marginal way to augment the price-induced conservation that appears quite dramatically in the statistics. High overall prices, coupled with selected grants to low-income families, schools, and hospitals to improve the energy efficiency of their buildings, will be a suitable mix of policies in the future. Some hard thinking needs to go into overcoming institutional barriers and into the implementation of conservation incentives; and a tough political decision must be made on pricing. There is no reason why a substantial gasoline tax should not be imposed, and imposed quickly. Gasoline-powered transportation is a luxury bestowed on few of earth's citizens, and those who have it ought to pay for it. The proceeds of a gasoline tax, which can be tremendous, could go to improving transportation-system designs and ultimately work to give everybody higher-quality transportation than would be the case without the tax.

The present patchwork of policies to induce coal to enter into markets and oil to leave those same markets, have been argued by some analysts to be counterproductive. This could well be so. For example, the law that prohibits oil and gas use in new power plants is moot. No power company would order an oil or gas plant under today's economic conditions. Instead of dictating who should or should not use what kind of fuel, the way could be opened up for greater coal production, transportation, and use. Mobilizing resources for the building of much larger coal ports on the United States' east and west coasts, faster coal leasing, and strict enforcement of environmental and pollution controls without years of litigation would induce coal producers to produce the coal, pay for the high environmental costs, and pass on the costs to buyers who would pay a total price still less than equivalent oil or gas prices.

Vulnerability to sudden oil-supply disruption increasingly worries U.S. energy-policymakers. The highly visible energy vulnerability concerns all U.S. policymakers. The reduction of United States oil-import dependence—or the elimination of vulnerability to supply disruption—is the basis for virtually every energy-policy proposal.

Of course, European countries and Japan also import oil. Yet these countries are not called on to provide military might for the Western world, nor are international oil prices yet denominated in their national currencies. For these reasons, and because the United States is also an oil producer, oil imports to the United States are a substantially different thing from oil imports to Europe and Japan.

Despite the need for change, much has been done in the last two or three years of the 1970s. Domestic oil prices have been decontrolled and the proceeds from the sudden increase in revenues will not all go to the oil companies, which at any rate are having difficulty investing all the new revenues they are earning. A spate of new initiatives to induce conservation and to protect those who need protection from higher prices have been initiated and a variety of incentives have been offered for the production of alcohol fuels, synthetic fuels, and the use of high-cost solar technologies.

These things have come about in spite of the political and institutional system of the United States. Some substantial credit must be given to President Carter. It is hoped that the mood of change sweeping across America at the beginning of the 1980s will not result in the undoing of all that has gone on before in energy policy.

International Energy-Policy Imperatives

U.S. foreign policy must recognize with greatly increased sensitivity the legitimate goals, aspirations, and complaints of the Third World. Many of the developing oil-importing countries have oil and gas resources that need to be more fully explored. At today's prices, marginal fields become economic. It is in the U.S. self-interest, and certainly in the global interest, to use every avenue to encourage the development of energy resources in developing countries and to help them in their development aims.

The World Bank has initiated a new program designed to address the needs of the oil-importing developing countries and is proposing to expand this to a level of $25 billion over the next five years. The ability of the World Bank to mount such a program is dependent on its obtaining additional capital, possibly through the creation of a separately capitalized energy affiliate. This high-priority item deserves complete United States support. Although the $25 billion from the World Bank for developing countries' energy investments would undoubtedly generate several times that amount in investment from other sources, more would still be needed. The cost to the energy-deficient countries of not being able to obtain the requisite capital could be staggering. They would either face bankruptcy and the collapse of their economies, or the prospect of becoming permanent wards of the international community. U.S. energy policy in the 1980s must recognize the massive requirements of capital, management skills, and technology that are indispensible ingredients of energy development in the developing world. If these needs are not met, there is the ever-present threat to our own national-energy needs.

Additional forums for candid discussion of North-South, producer-consumer issues are needed. Indeed, there may be no more important issue on the global agenda—or among the energy policy priorities in the United States—than dialogue with developing countries to seek solutions to their severe development problems, in which energy plays a major role. The United States must find a way to sort out its foreign-policy dilemmas; blind and unquestioning support of Israel in the face of strong Arab feelings on the Palestinian issue and the rights of homeland simply cannot continue.

The International Energy Agency provides a useful forum for dialogue between America and its Western allies, and its progress in encouraging energy conservation and in developing a workable oil-shortage sharing scheme is commendable. One would hope, however, that the IEA posture could evolve to less of a confrontational "consumers cartel" and more toward an agency potentially cooperative with producers—that is, to instigate a producer-consumer dialogue.

The Third World countries have legitimate aspirations and legitimate complaints. They want, and should get, a fair return on their investments. They want, and it should be recognized that they have, a valuable resource, too precious to be simply burned up by extravagant American and European cars. The non-oil-importing countries, strangled by trade patterns favoring the developed North, and by highly fluctuating commodity prices on which nearly all of them so deeply depend, need a sensitive awareness of their plight on the part of the developed countries who possess the wealth, economic power, and integrity necessary to find solutions which will, in the end, benefit us all. For the pragmatist must realize that nearly all of the growth in markets for new products will be in developing countries in the future; if developing countries do not make it, then industrial countries do not make it either.

Epilogue: Changes under President Reagan

Seven months into the administration of President Ronald Reagan, evidence appears to support the broad generalizations made in this chapter regarding new slants in U.S. energy policy.

President Reagan, as has been widely publicized, succeeded dramatically in his first few months in office in overpowering Congress, in taking the initiative in major issues and in getting his way. It is almost as if he formed a majority in the parliamentary sense, and could therefore feel confident that he was truly at the helm of a single-purpose government. It may prove to be so. If it does, it represents a major coup in American politics. However, the President's dominance on budget and tax issues of national scope will probably not persist when the issues addressed become more regional and issue-specific.

Congress still holds the balance of power on such narrower policy issues. Regional interests and special interests can soon regroup and rally Congress to their causes, with the usual result that decisionmaking will slow and compromises will be forced.

The new administration appears little concerned with energy policy per se. Whereas earlier the lack of coherent energy policy in the United States was decried, now it is seen to be a virtue. The right energy policy, it is now believed, is for government to stay out of the energy business (with the notable exception of nuclear power) as much as possible. The new administration emphasizes market mechanisms. It focuses on domestic energy production, and lacks enthusiasm for energy conservation as a cause aside from the conservation associated with deregulated energy prices. It promotes a reversal in the trend of an ever-increasing role of government in energy affairs.

The new trend manifested itself in rather surprising and sweeping budget cuts—more specifically offered as part of the new supply-side economics than as overt energy policy. The cuts struck at every energy program rather forcibly, with three interesting exceptions: the gasohol program, nuclear power, and a strategic petroleum reserve. Funds for the latter have, at this writing, actually been stricken from the budget, but commitment remains for a major government role through off-budget funding. Gasohol enjoys bipartisan popularity, in spite of questionable merits.

The increase in funding for nuclear energy, and particularly for breeder-reactor research, is curious, given the heavy cuts in virtually all other programs: Nuclear energy can offer only base-load electricity to the U.S. energy mix—roughly 10 percent of the total supply. And various studies indicate that nuclear power—and breeder-reactor power particularly—may not be economically competitive. The decisiveness of the new administration, contrasted with the earlier tendency to keep all the options open for future energy supply, is laudable. But why should most promising and dubious supply (and demand) options be required to survive the test of the market system, while the most expensive, rather narrow source (nuclear power) is spared?

The Reagan Administration has shifted emphasis to long-run research and away from commercialization. It has accelerated leasing of federal lands for oil exploration. It has speeded decontrol of crude oil and product prices. It has ended or drastically cut many energy programs through the budget process, particularly the synthetic-fuels program, and initiated no new ones. It has seemed to downplay the previously overriding concern about U.S. dependence on oil imports, and abandoned any talk of energy goals. In short, it has undone quite a lot of energy policy in a short time. It has done this, it says, in order to unleash the vast energy resources it sees in America, ready to be exploited by a dynamic and innovative economic system (the oil companies, one would presume) if only government will get out of the way.

While it is much too early to tell if this philosophy will prove valid, certain trends seem likely to prevent much opposition. Exploratory drilling rates in the United States are high—breaking records every week; oil will presumably be found and the decrease in U.S. oil output may be slowed. Meanwhile, oil use continues to fall in the United States and the industrial world. The current glut of world oil seems likely to persist unless there are drastic upheavals among overseas suppliers. These trends make it difficult to find motivation for new programs of conservation or government involvement on the supply side.

The drop-off in oil demand is likely to last—more because of structural responses to the 1971-1974 price increases than to the current recession. But government incentives to shift away from oil use at an even faster pace, and standards to force improved efficiency have been effective, and are needed in many sectors where the market does not adequately reflect the real price of energy.

The pervading philosophy seems certain to ignore the potential multiple virtues of an energy tax. Taxes on energy reflecting its true social and economic costs can promote faster shifts to energy-efficient processes and away from oil, and can generate enormous revenues for budget-balancing purposes. For example, a $1.20-per-gallon tax on gasoline in 1978, at 1978 gasoline-use levels, would have generated revenues equal to *all* corporate income taxes plus one-half of all personal income taxes.

The United States appears to be engaged in an experiment in market mechanisms as a cure for all problems. The shift from past policy thereby could be quite drastic. The results of the experiment are unpredictable; but if its philosophy is not too narrow or rigid, opportunities for action will not be missed and the experiment could well be a satisfying success.

Notes

1. The author gratefully acknowledges the assistance of Michael T. Woo and Jeff Greene of the staff of the Subcommittee on Energy and Power, U.S. House of Representatives, who provided information used in the preparation of this chapter.

2. Senate Foreign Relations Committee, Subcommittee on Multinational Corporations, *Report on Multinational Corporations and U.S. Foreign Policy*, 93rd Cong., 2nd Sess. (Washington, D.C.: 1975). This followed the earlier major report of the Federal Trade Commission to the Senate Small Business Committee, *The International Petroleum Cartel*, 82nd Cong., 2nd Sess. (Washington, D.C.: 1952).

3. Quoted in Subcommittee Print, U.S. Senate Subcommittee on Multinational Corporations, *The International Petroleum Cartel, the Iranian Consortium, and U.S. National Security* (Washington, D.C.: 1974).

4. Ibid.

5. Lloyd M. Cutler, "To Form a Government," *Foreign Affairs* 59:1 (Fall 1980), pp. 135-136. Reprinted with permission.

References

Cutler, Lloyd. "To Form a Government," *Foreign Affairs* (Fall 1980).

Landsberg, Hans H., ed. *Energy: The Next Twenty Years.* Report of a Ford Foundation Study administered by Resources for the Future. Cambridge, Mass.: Ballinger, 1979.

U.S., Congressional Budget Office. "The World Oil Market in the 1980's: Implications for the United States," Report. Washington, D.C.: 1979.

U.S., Department of Energy. *National Energy Plan II*, A Report to the Congress. Washington, D.C.: 1979.

______. *Secretary's Annual Report to the Congress.* Washington, D.C.: 1980.

______. *Securing America's Energy Future: The National Energy Policy Plan.* A Report to the Congress. Washington, D.C.: July, 1981.

Zausner, Eric. *The United States Energy Future: Difficult Choices in Difficult Times.* Bethesda: Booz-Allen-Hamilton, Inc., 1980.

13 Canadian Energy Policy

John F. Helliwell

Canada is unique among the energy-rich industrial countries in the extent to which its energy resources are under the ownership and control of its provincial governments. This has meant that national energy policy, especially as it relates to crude oil and natural gas, has after each oil crisis been more concerned with the sharing of the windfall gains than with efficiency of energy use or acceleration of plans to provide domestic substitutes for imported oil. Seen in this context, the controversial federal National Energy Program of October 1980 is less a new departure than the type of response that is to be expected when sharp changes in world energy prices make unworkable the energy pricing and taxation compromises that were worked out under different circumstances. It is probably true, however, that the current federal policy, and the responses to it by the main producing provinces, reflect a greater departure from negotiated compromise than ever before. There is thus more risk of a prolonged policy stalemate that will lessen Canada's political cohesiveness and lessen the degree of energy self-sufficiency likely to be achieved during the 1980s. In the survey that follows, enough background to the Canadian energy situation will be provided to place the current controversies in a broader context, and hence to provide some basis for guessing the likely outcomes for Canada, and the implications for the other industrial countries of North America and Western Europe.

Energy Production, Energy Trade, and Energy Use

Energy Production and Trade in Canada and the United States

Tables 13-1 and 13-2, based on data and forecasts published by the International Energy Agency (IEA), show energy production and net energy imports, and, by inference, consumption in both Canada and the United States of the five main sources of primary energy: crude oil, natural gas, coal, nuclear electricity, and hydroelectric power. Actual data are reported for 1960, 1973, and 1976, and IEA forecasts (based on 1977 submissions to IEA by the federal governments of both countries) for 1980, 1985, and 1990. The table 13-1 figures for oil and gas production suggest, and data for

Table 13-1
Energy Production in Canada and the United States
millions of metric tons of oil equivalent (mtoe)

Country/Type of Power[b]	*Year*					
	1960	*1973*	*1976*	*1980*[a]	*1985*[a]	*1990*[a]
Canada						
Oil	26.4	99.5	81.1	81.3	76.1	89.7
Gas	11.6	62.1	57.4	67.7	73.2	66.3
Coal	5.8	12.0	14.1	21.3	29.9	37.1
Nuclear	—	3.9	4.4	8.9	16.9	29.3
Hydroelectric	33.5	49.7	57.2	61.3	73.4	87.4
Total	77.3	227.2	214.2	240.5	269.5	309.8
United States						
Oil	382.8	511.2	454.8	517.8	516.1	457.0
Gas	293.8	514.5	453.6	385.1	379.4	386.1
Coal	251.9	346.8	388.8	506.0	613.0	719.0
Nuclear	0.1	20.6	46.3	86.0	187.3	292.2
Hydroelectric/other	36.3	67.8	73.9	80.0	82.4	97.6
Total	964.9	1,460.9	1,417.4	1,574.9	1,778.2	1,951.9

Source: International Energy Agency, *IEA Reviews of National Energy Programmes* (Paris: OECD, 1977). All data are annual f'ows measured in millions of net metric tons of oil equivalent (mtoe).

[a]IEA forecast production.

[b]Conversion factors: crude oil, 1 million barrels per day = 50.35 mtoe for Canada, 49.1 mtoe for the United States; gas, 1 trillion cubic feet = 23.11 mtoe for Canada, 23.89 mtoe for the United States; coal, 1 million metric tons = 0.69 mtoe for hard coal, 0.34 mtoe for lignite; electricity, 1 terra watt hour = .086 mtoe.

exploration and reserves confirm, that Canadian oil and gas reserves have been developed later than their United States counterparts. In 1960, before the Ottawa Valley Line was established (in 1961, following the recommendation of the Borden Commission) to provide a protected market west of the Ottawa Valley for Canadian crude oil, Canada had net oil imports more than half as large as domestic oil production. By comparison, U.S. net oil imports in 1960 were less than 20 percent of domestic production. With the protection of the Ottawa Valley Line, and with increasingly easy access to U.S. markets in the late 1960s and early 1970s, Canadian oil production increased almost fourfold between 1960 and the mid-1970s.

Natural-gas and coal production have been historically much more important in the United States than in Canada, chiefly because there were large markets close to large deposits. Indeed, the Canadian steel industry, centered in southern Ontario, and thermal production of electricity by Ontario Hydro, are heavily dependent on coal imported from nearby U.S. deposits. When Canadian natural-gas production eventually started to grow dramatically—it increased more than fivefold between 1960 and 1973—

Table 13-2
Net Imports of Energy into Canada and the United States
millions of metric tons of oil equivalent (mtoe)

Country/Type of Power[b]	*Year*					
	1960	*1973*	*1976*	*1980*[a]	*1985*[a]	*1990*[a]
Canada						
Oil	14.5	−11.5	10.1	19.1	34.1	31.5
Gas	−2.4	−23.7	−21.9	−23.6	−20.5	−5.1
Coal	7.2	3.4	2.0	0.7	−3.1	−5.0
Total	19.3	−31.8	−9.8	−3.8	10.5	21.4
United States						
Oil	70.0	284.5	353.3	506.3	571.6	696.3
Gas	0.3	12.2	18.3	33.9	54.2	52.3
Coal	−21.4	−20.6	−38.7	−50.0	−59.0	−62.0
Total	48.9	276.1	332.9	490.2	566.8	686.6

Source: International Energy Agency, *IEA Reviews of National Energy Programmes* (Paris: OECD, 1977).

[a]IEA forecast figures.

[b]Conversion factors: crude oil, 1 million barrels per day = 50.35 mtoe for Canada, 49.1 mtoe for the United States; gas, 1 trillion cubic feet = 23.11 mtoe for Canada, 23.89 mtoe for the United States; coal, 1 million metric tons = 0.69 mtoe for hard coal, 0.34 for lignite; electricity, 1 Terra Watt Hour = .086 mtoe.

the impetus came partly from the increasing stocks of proven gas reserves discovered in the search for oil but primarily from the development of pipeline transmission systems designed to serve U.S. as well as Canadian markets. In a similar way, the forecast expansion of Canadian coal production by 175 percent between 1976 and 1990 (the actual increase is likely to be smaller) is in large measure based on the shipping of western-Canadian metallurgical coal to Japanese and Korean markets, along with greater use of western-Canadian coal in Ontario.

Hydroelectricity has been, and will continue to be, of about equal absolute size in Canada and in the United States, and hence of far greater relative importance in the much smaller Canadian economy (about 10 percent of U.S. figures in both population and GNP). Relative to total energy requirements, nuclear energy has played similar roles in both countries, with 1976 and forecast 1990 nuclear capacities in the United States about ten times as large as in Canada.

In the forecast period, Canada's natural-gas exports are assumed to taper off as existing contracts expire in the late 1980s and early 1990s. The figures thus ignore the 3.75 trillion cubic feet (tcf) of new gas exports approved in December 1979 for delivery to the United States between 1980 and 1987. Additional exports beyond 1987 are also likely, but have not yet been approved. Total primary-energy demand is forecast to grow by 38 percent between 1976 and 1985, and oil production is forecast to fall, thus causing forecast oil imports to reach about .7 million barrels per day (mbd) in 1985. The oil-supply forecast appears to involve about 1.5 mbd, of which more than .4 mdb is presumed to come from oil sands and heavy-oil plants. The 1985 and 1990 estimates of natural-gas production appear to involve some production from frontier deposits (Mackenzie Delta, Arctic Islands, or east-coast offshore), amounting to 2 percent of total production in 1985 and 15 percent in 1990. The 1985 and 1990 forecasts of conventional oil production (estimated to be down to .7 mbd in 1990) and of nonfrontier natural-gas production are intended to represent maximum sustainable production from nonfrontier sources. As I shall indicate later, these are serious underestimates of what nonfrontier suppliers will be willing and able to produce. Thus the table 13-1 figures are likely to overestimate oil imports, to underestimate gas exports, and to overestimate production of oil sands and very heavy oils.

The forecasts in the table 13-1 were prepared before the doubling of world oil prices in 1979. To the extent that these price increases eventually show up in prices for Canadian energy production and consumption, future energy demands will be smaller, and some types of domestic energy production larger, than shown in table 13-1.

International Comparisons of Primary-Energy Use

Tables 13-3 and 13-4 show total primary-energy use, per capita and in relation to real Gross National Product (GNP), for Canada and the United States, along with comparable measures for other OECD countries. Sweden and Norway were selected because like Canada they have northern climates, high standards of living, and heavy energy use associated with their forest industries. New Zealand has a slightly milder climate, but shares with Canada and the Nordic countries a heavy use of hydroelectricity (and also the presence of electricity-intensive aluminum smelting), and a substantial forest industry. Canada is more sparsely populated than the other countries, with 2 inhabitants per square kilometer, compared with 12 for Norway and New Zealand, 20 for Sweden, 23 for the United States, 247 for Germany, and 299 for Japan. The latter two countries are included for comparison to show the effects that their rapid industrial growth has had on energy-use patterns.

All of the countries show past and future increases in the per-capita use of primary energy. Canada's average annual increases (in absolute terms) are larger than those for all of the other countries, for past as well as future periods. In terms of the ratio of energy to output of goods and services, most countries show constant or increasing energy use between 1960 and 1973, a period during which the relative price of energy fell in all countries. From 1973 to 1976, most of the countries showed falling energy-intensity

Table 13-3
Total Primary-Energy Use Per Capita in Canada and Selected Other OECD Countries
millions of metric tons of oil equivalent (mtoe)

Country	Year					
	1960	*1973*	*1976*	*1980*	*1985*	*1990*
Canada	5.36	8.78	8.87	9.78	10.94	12.27
United States	5.60	8.30	8.10	9.20	10.00	10.70
United Kingdom	3.26	4.00	3.70	4.02	4.29	4.62
Sweden	3.64	5.79	6.05	6.40	7.21	7.24
Norway	2.50	4.80	5.05	5.76	6.55	7.20
New Zealand	2.29	3.27	3.48	3.87	4.67	5.20
Germany	2.63	4.37	4.21	5.03	5.70	6.40
Japan	1.02	3.15	3.24	3.91	4.88	5.70

Source: International Energy Agency, *IEA Reviews of National Energy Programmes* (Paris: OECD, 1977).

of output, and from 1976 to 1990 all countries except New Zealand forecast substantial declines in energy intensity.

The forecast drop in the ratio of energy use to output is of the same order for Canada as for most of the other countries. What is somewhat surprising is that the gap between Canada and other countries actually widens, in proportionate terms, from 1976 to 1990. To the extent that Canada's past energy intensity was because of low-cost and low-priced energy, the gap should be closed as prices move toward world levels, as they are assumed to do in the forecast assumptions underlying tables 13-3 and 13-4.

Table 13-4
Total Primary-Energy Use in Relation to GNP (Canada and the United States) or GDP (Other Countries)
mtoe/billions of U.S. dollars (1970)

Country	Year					
	1960	*1973*	*1976*	*1980*	*1985*	*1990*
Canada	1.93	1.96	1.87	1.80	1.71	1.64
United States	1.51	1.51	1.51	1.41	1.41	1.37
United Kingdom	1.85	1.67	1.52	1.41	1.32	1.24
Sweden	1.30	1.34	1.32	1.19	1.16	1.09
Norway	1.31	1.54	1.41	1.24	1.20	1.12
New Zealand	1.26	1.36	1.40	1.51	1.67	1.69
Germany	1.25	1.27	1.20	1.23	1.11	1.04
Japan	1.33	1.33	1.34	1.32	1.28	1.26

Source: International Energy Agency, *IEA Reviews of National Energy Programmes* (Paris: OECD, 1977).

International Comparisons of Electricity Use

The high and rising per-capita energy-consumption figures for Canada require further investigation and explanation. It is useful to consider electricity separately from other forms of final energy, if only because of the difficulty of preparing and interpreting primary-energy statistics for electricity. The problem is one of finding a suitable oil-equivalent measure of the electricity obtained from nuclear, hydroelectric, and geothermal power. The IEA procedure is to use average efficiencies of thermal generation facilities in each country or in some of its neighboring countries. For countries such as Canada, Sweden, and Norway that have had large and low-cost hydroelectric sites (often far from major centers of population) and have attracted large electricity-intensive industries (such as aluminium smelting) as a consequence, the use of oil-equivalent measures based on thermal generation may exaggerate the figures for total energy use. To isolate the impact that earlier availability of low-cost hydroelectric sites may have on current patterns of electricity use and total energy use, table 13-5 shows per-capita electricity use, the relative importance of electricity as a fraction of total primary energy, and total primary-energy use excluding electricity. The table shows that Canada, Norway, and Sweden, all predominantly hydroelectric-countries, use much greater than average amounts of electricity, especially for industrial use. New Zealand, the other predominantly hydroelectric-country, shows something of the same pattern for households, but the corresponding pattern for industrial use is not evident in the 1976 data even though about 15 percent of New Zealand's electricity was being consumed by the country's new aluminium smelter. The rising energy/GDP ratios forecast for New Zealand in table 13-4 imply further shifts to energy-intensive industry, to an extent that seems implausible.

Even after all electricity is eliminated from primary-energy-use figures, shown in the last column of table 13-5, the figures still show Canada's primary-energy use to be as high as that of the United States, on a per-capita basis, and much higher than that of all sixteen other countries. If similar calculations are done for 1990, they show per-capita nonelectricity energy-use ratios of 7.1 for Canada compared with 5.8 for the United States, and much less for all other countries. The calculations excluding electricity naturally understate the energy-intensiveness of electricity-intensive countries like Canada, yet even in this case the official forecasts show that Canada is expected to be increasingly more energy-intensive than other countries. The calculations that underlie this forecast almost surely understate the long-term effects of energy conservation if Canadian energy prices and costs rise until Canadian energy-users face roughly the same energy costs as in other countries.

So much for the broad patterns of national energy supply and demand. To understand better the workings and effects of Canadian energy policies,

Table 13-5
1976 Electricity Use in Canada and Other OECD Countries
(thousand kWh per capita)

Country	*Population mid-1976 (millions)*	*Household Use*	*Industrial Use*	*All Uses*	*Electricity as Percentage of TPE*[a]	*TPE Per Capita Excluding Electricity*
Canada	23,143	2.98	5.14	12.28	38	5.43
United States	215,118	2.80	3.35	9.91	32	5.51
United Kingdom	56,001	1.52	1.79	4.60	34	2.44
Sweden	8,219	2.42	4.97	10.52	40	3.62
Norway	4,027	4.84	10.03	18.62	59	2.13
New Zealand	3,116	2.69	2.25	6.71	44	1.94
Germany	61,513	1.17	2.61	5.11	31	2.94
Japan	112,768	0.87	2.71	4.34	20	2.17

Source: International Energy Agency, *The Electricity Supply Industry in OECD Countries* (Paris: OECD, 1978).

it is essential to know something of the regional distribution of energy resources and energy demands. In the simplest terms, it is enough to know that Alberta accounts for about 85 percent of Canada's oil production (Saskatchewan produces most of the rest) and 85 percent of natural-gas production (British Columbia produces most of the rest). Canada's coal production is 42 percent in Alberta, 31 percent in British Columbia, 19 percent in Saskatchewan. Thus Canada's hydrocarbon production is almost entirely from the three most westerly provinces. The major centers of population, industry, and energy use are in Ontario and Quebec, although energy use is much more evenly spread than is energy production. Ontario accounts for 31.4 percent of Canada's oil consumption, 46.6 percent of natural-gas consumption, and 51 percent of coal consumption. Quebec, which until 1976 received all of its crude oil from offshore sources and is only partially served by natural-gas transmission systems, accounts for 29.7 percent of oil consumption, 5.5 percent of natural-gas consumption, and 2 percent of coal consumption. The east-coast Maritime provinces account for 13.2 percent of Canada's oil consumption (all imported), consume no natural gas, and 3 percent of coal. They account for 7 percent of Canada's coal production. East-coast offshore oil and gas drilling has so far led to some discoveries but no production.

Although the distribution of hydroelectric installations is somewhat uneven across the country, with Quebec and British Columbia having the major resources, all of the provinces generate the balance of their own requirements from nuclear power (in Ontario, and planned for Quebec) and from conventional thermal installations. Thus there is little net interprovincial or international trade in electricity, beyond the large block of power (almost 10 percent of Canadian production) going from Labrador to Quebec on long-term contract.

Energy Pricing and Revenues

Federal Energy-Pricing Policies

The federal government came by its strategy for oil and gas pricing in bits and pieces, starting with a fortuitous timing of events in 1973. Exports of crude oil came under direct control under the existing powers of the National Energy Board (NEB) in March of 1973, in response to concern that the rapid growth of export demands (an 85 percent increase from 1970 to 1973) might deprive Canadian refineries of their crude-oil supplies. In September 1973, in response to consumer complaints about higher oil prices unmatched by rising costs, domestic oil prices were frozen at 3.80 Canadian dollars per barrel, and a special charge levied on exports, one month before the war in the Middle East and the ensuing oil embargo.

Then world oil prices started to climb dramatically, and the other main elements of the Canadian pricing policy fell into place in response to distortions coming about when the oil-price controls really started to take effect. Thus in January 1974 it was agreed between federal and provincial governments that oil should have the same price (except for transport costs) throughout the country, and an import-compensation scheme was developed to subsidize refineries using higher-cost imported oil. The pressure to raise natural-gas prices came primarily from the provincial governments in Alberta and British Columbia.

Preemptive action by British Columbia and pressure from Alberta forced the federal government to act, through the NEB, to set gas-export prices closer to those for crude oil. Table 13-6 shows the main policy-determined increases in domestic wellhead prices for oil, Toronto city-gate prices for natural gas, and natural-gas export prices.

Regional Distribution of Oil and Gas Use and Revenues

In regard to the broader aspects of the effects of energy prices on revenue, it is difficult to make a complete accounting but easy to show the striking size of the changes in Canadian regional income flows that have resulted from energy-price increases since 1973. It is necessary to consider only oil and gas, because, as already indicated, there is little interprovincial trade in electricity.

In 1980, the world value of Canada's 500 million barrels of oil production was about 20 billion Canadian dollars, and the 2.5 trillion cubic feet of gas production had an export value of about $13 billion. (Unless otherwise mentioned, all amounts are in Canadian dollars, $1 Canadian being worth about $.85 U.S.) This amounts to about 12 percent of Canada's 1980 GNP. Alberta received an average price of about $15.75 per barrel (about 40 percent of the

Table 13-6
Crude Oil and Natural Gas Prices, 1973-1981
(Canadian dollars)

Date when New Prices Were Established	Crude-Oil Wellhead Price Per Barrel	Natural Gas Prices Per Million Cubic Feet: Toronto City-Gate Wholesale Price[a]	Natural Gas Prices Per Million Cubic Feet: Export Price
September 1973	3.80		
April 1, 1974	6.50	0.62	
November 1, 1974		0.82	
January 1, 1975			1.00
July 1, 1975	8.00		
August 1, 1975			1.40
November 1, 1975		1.25	1.60
July 1, 1976	9.05	1.405	
September 10, 1976			1.80
January 1, 1977	9.75	1.505	1.94
July 1, 1977	10.75		
August 1, 1977		1.68	
September 20, 1977			2.16[b]
January 1, 1978	11.75		
February 1, 1978		1.85	
July 1, 1978	12.75		
August 1, 1978		2.00	
May 1, 1979			2.30[b]
July 1, 1979	13.75		
August 1, 1979		2.15	2.80[b]
November 4, 1979			3.45[b]
January 1, 1980	14.75		
February 1, 1980		2.30	
February 17, 1980			4.47[b]
August 1, 1980	16.75		
September 1, 1980		2.60	
January 1, 1981	17.75		
April 1, 1981			4.94[b]
July 1, 1981	18.75		

[a]The natural gas prices do not include the federal natural gas gax introduced in October 1980. This tax was $.30/mcf from November 1, 1980, $.45/mcf from July 1, 1981, $.60/mcf from January 1, 1982, and $.75/mcf from January 1, 1983.

[b]Beginning with the September 1977 export price increase, export prices of natural gas are listed in U.S. rather than Canadian dollars.

landed price of offshore crude oil for the 450 million barrels of oil exported from the province and an average border price of about $2.50 per million cubic feet (mcf), about half of the export value, for the 1.8 tcf of gas exported from the province. Even with energy prices much below world prices, Alberta's gross revenues of almost $12 billion in 1980 were large by any standard, whether in relation to population (Alberta had 9 percent of Canada's mid-1980 population of 24 million, so 1980 oil and gas revenues amounted to about $6,000 per capita) or in relation to past experience. In 1980, the Alberta government received about $5 billion, or $2,500 per capita, in production royalties and land payments for oil and natural gas.

The large size and rapid growth of actual and potential oil and gas revenues explain why their distribution among producing firms, governments, and consumers has been such an active source of debate, legislation, and jurisprudence in Canada since 1973. The intricacies of the conflicts and compromises in taxation, regulation, and development are much too involved to unravel in a brief survey. There are many analyses available, although none has yet appeared with the full objectivity and clarity that only hindsight can bring. This survey indicates only the key features of the fiscal debate as part of a general description of the interests and investments of the main participants.

Energy-Policy Actors and Instruments

The first and most obvious point to make is that oil and gas have much in common with each other, but little in common with electricity and coal. Nuclear energy is in yet a different category. The fuels will be treated separately, starting with oil and gas.

Oil and Gas

Oil and gas resources in Canada are almost entirely owned by the Crown, and developed under lease by predominantly foreign-owned firms. Natural resources fall within provincial jurisdiction under the British North America Act, so that the *Crown* for these purposes means the provinces south of 60 degrees latitude, and the federal government for resource deposits in the Yukon and Northwest territories. Under the National Energy Program (NEP) introduced by the federal government in October 1980, all offshore areas and lands north of 60 degrees were described as the *Canada lands* and made subject to direct federal control, higher levels of federal taxation, and higher levels of subsidy for oil and gas exploration for firms with a substantial degree of Canadian ownership.

Federal Fiscal Tools. For the oil and gas resources that lie within the provinces, the chief federal fiscal instrument prior to October 1980 was the federal corporate income tax. The key changes that have been made in the federal corporate income tax since 1973 started with the disallowance of royalty payments to provinces as a deductible expense in 1974, and continued with a number of measures designed to refund or defer income tax for income reinvested. The federal budget of October 1980 introduced two major new tax instruments: an 8 percent tax on net wellhead revenues from petroleum and natural gas, and a commodity tax on all natural gas, whether used in Canada or exported. The latter tax started at $.30 per mcf in November 1980 (February 1981 for export gas) and is intended to reach $.75 per mcf in 1983.

Provincial Fiscal Tools. The main provincial fiscal instrument for oil and gas is the production royalty, although land sales and bonus bids for production rights can be very important sources of revenue, especially in the fiscal regime prior to October 1980, in which taxes and royalties were structured to favor reinvestment of current revenues.

The Constitutional Issue. The key constitutional issue of the mid-1970s (especially in Saskatchewan) relates to provincial attempts to design production royalties that are flexible enough to permit firms to recover the costs of producing from high-cost deposits while allowing the Crown to collect a high proportion of the resource value from low-cost deposits. In a controversial decision, the Supreme Court of Canada decided in favor of firms that were appealing the Saskatchewan legislation on the grounds that it amounted to an indirect tax. Under the British North America Act, both federal and provincial governments may levy direct taxes, but only the federal government may levy indirect taxes. The provinces also levy a corporate income tax which for most provinces including the three main oil-and-gas producing provinces is collected by the federal government. The provinces continued to let royalties be deductible for their own corporate income taxes, and also offered some compensation for extra federal income tax incurred because of the nondeductibility of provincial royalties. The provinces continue to argue that the federal action was an unjustified and discriminatory infringement on the provincial ownership of natural resources.

The Federal Government as Oil and Gas Developer. The main federal investments in energy production include: almost 50 percent shareholding in Panarctic Oils Limited, the main explorer in the Arctic islands and the driving force behind the Polar Gas pipeline proposal; 100 percent ownership of Atomic Energy of Canada Limited; 15 percent equity in the Syncrude oil-sands project; and other energy investments undertaken by Petro-Canada Limited, formed in 1976 as the federal government's wholly-owned vehicle for direct investment in oil and gas development. In November 1978, Petro-Canada made by far its largest expenditure, $671 million (U.S.) to acquire Phillips Petroleum's 48-percent controlling interest in Pacific Petroleums Limited of Calgary, which in 1976 was Canada's fourth-largest producer of natural gas and ninth-largest producer of oil. Subsequent purchases raised Petro-Canada's holding to 100 percent. It is estimated, based on 1977 data, that with the amalgamation of Pacific Petroleums' operations, Petro-Canada's production would be Canada's sixth-largest in crude oil (including natural-gas liquids) and second-largest in natural gas. The continuing controversy surrounding Petro-Canada's purchase involves issues of public versus private ownership, foreign versus domestic control, and new capital investment versus takeovers as means of establishing a presence. The problems raised by the further blurring of federal government roles have not received much public attention, but are of potential importance. For example, Petro-Canada, through its new control of Pacific Petroleums, is now the dominant shareholder in Westcoast Transmission, a federally regulated company that is a 50-percent partner in the Alaska Highway natural-gas pipeline. In the National Energy Program of October 1980, the federal government announced that it intends to acquire several of the large oil and gas firms, with Petro-Canada as the agent of acquisition. The document proposes that Petro-Canada might keep some small part of the assets to round out its activities, "while the remaining assets will form the basis for one or more new Crown corporations."

A final and most important class of federal instrument is based on the federal power to tax and regulate international trade. These powers are of special importance in the case of oil and gas, in which international trade has been such an important and controversial part of the industry's development. The National Energy Board is the chief federal agency overseeing energy trade and development; its approval of energy exports and pipeline projects is a necessary but not sufficient condition for federal-government approval, because proposals must also receive cabinet approval. The NEB also recommends to the government the levels of the natural-gas export price and the oil-export charge described in the previous section.

Provincial Governments as Oil and Gas Owners and Developers. The provincial governments start with resource ownership as their primary policy instrument in oil and gas development, but there have been three other main types of policy instruments. First, Alberta has long controlled the export of oil and gas from the province through the hearings and decisions of the Alberta Energy Resources Conservation Board (AERCB). Second, prices for gas production have been forced up by means of changes in the Arbitration Act, and prices for Alberta users have been lowered by means of rebates paid to natural-gas users by the Alberta government. In British Columbia, the wholesale prices for B.C. users of natural gas have been kept low (less than one-quarter of the export price at the end of 1980) by having the Crown-owned British Columbia Petroleum Corporation buy all gas from

B.C. producers and sell at one price in the domestic market and another price in the export market. Outside British Columbia and Alberta, natural-gas prices to Canadian users are based on the Toronto city-gate price, which was, prior to October 1980, set by negotiation between the federal and provincial governments. Third, the provincial governments have been directly involved in numerous corporate ventures intended to support provincial revenue or development goals.

The Oil and Gas Industry. The chief participants in the oil and gas industry are the firms that discover, develop, process, produce, transport, and distribute the oil and gas. The oil industry is vertically integrated from exploration to the gasoline pump, with most of the crude-oil production and almost all of the refining and marketing activity in the hands of the major international oil companies. The natural-gas industry is much more dispersed, with a large group of firms (including the major international firms that dominate oil) in exploration and development, three large transport companies (AGTL, Westcoast Transmission, and TransCanada Pipelines) and a large number of provincially regulated natural-gas distribution companies.

The main policy instrument available to the producing firms is political and administrative influence, acquired through financial and personal contacts, and supported by privileged access to expertise and data, and enforced by actual or threatened mobility. By their mobility, the firms force competition among the resource-owning jurisdictions. Thus when the federal and provincial governments both moved strongly to increase oil and gas taxes and royalties in 1974, there was a sharp reduction in drilling activity and there were much-publicized but not very large movements of drilling rigs from Canada to the United States. Shortly afterward there were important tax and royalty concessions by both levels of government, and the pace of drilling activity soon increased to record levels. In the first months following the October 1980 federal budget, there have been more substantial and more widely publicized movements of drilling rigs from Canada to the United States. To the extent that they can suppress competition among themselves when they deal with governments, the producing firms could jointly have enough leverage to obtain supernormal returns. Assessing the evidence on this point is not easy, as the concept of supernormal returns (returns greater than those necessary to attract the necessary pool of capital, labor, and industry) is itself not easy to define, and must be measured over the whole producing life of the resource, a time long enough to permit the relative bargaining strengths of the producing firms to ebb and flow several times.

How Are the Economic Rents Being Divided? Table 13-7 shows the distribution of economic rents (the amount by which the market value of

Table 13-7
Net Economic Income from Nonfrontier Crude Oil and Natural Gas under Alternative Pricing-and-Taxation Policies
(billions of end-1980 Canadian dollars)

Energy-Price Assumptions	*Total Income*	*Oil and Gas Industry Income*	*Canadian Energy-Users Income*	*After Waste of*	*Federal Government Income*	*Provincial Governments' Income*	*Total Canadian Income*
Case 1 (Alberta) totals	529	79	146	43	43	242	453
Case 2 (Full budget) totals	519	25	212	133	76	188	480
Case 3 (Alberta response) totals	460	10	194	133	69	169	432
Case 4 (Compromise) totals	527	30	199	122	82	198	484

Source: J.F. Helliwell and R.N. McRae, "The National Energy Conflict," *Canadian Public Policy* (Winter 1981) table 1. The appendix to that article explains the assumptions in more detail.

the oil and gas exceed the opportunity costs of all the capital, labor, materials, and industry needed to develop them) over the past and future production from the (conservatively estimated stocks of) conventional oil and gas deposits in western Canada. The results are based on an assumed world oil price of about $41 (or $35 U.S.) per barrel in 1981, rising to the end of the century at 2 percent annually in real terms. The four cases in the table show the economic rents under three different assumptions about Canadian energy prices and taxation policies. All cases show the distribution of the present values of all past and future oil and gas revenues using world prices for crude oil to set opportunity costs, and using actual prices and tax policies to calculate the distribution of revenues up to and including 1980. For 1981 and beyond, four different policies are considered. The first case shows the distribution of costs and benefits with the tax and royalty system in effect in mid-1980, and following the crude-oil price increases proposed by the government of Alberta in July 1980. These involve wellhead price increases of approximately $5 per barrel in 1981 and 1982, and $7 per barrel in 1983, with subsequent increases to keep the price at 75 percent of the world level. It is assumed that no federal-tax changes take place.

The second case embodies the main elements of the national Energy Program of October 1980, which are:

1. Wellhead price increases for crude oil of $2 per barrel in 1981, 1982, and 1983, $4.50 per barrel in 1984 and 1985, and $7 per barrel thereafter until reaching the lower of 85 percent of the world price or a reference price of thirty-eight 1980 Canadian dollars. There are no wellhead price increases for natural gas in 1981, and all subsequent increases are on a btu-equivalent basis with the crude-oil price increases, $.15/mcf for each $1 per barrel.
2. A petroleum-compensation charge on all users of oil products. This charge is $5.05 per barrel in 1981, $7.55 per barrel in 1982, $10.05 per barrel in 1983, and is assumed thereafter to be set large enough to cover in each year the costs of the subsidy payments on imported and synthetic oil.
3. The transfer of 50 percent of the revenues of the oil-export tax to the producing provinces.
4. The new natural-gas and liquids tax, starting at $.30/mcf and rising to $.75/mcf by 1983.
5. The 8-percent petroleum and gas revenue tax.
6. The phasing out of the depletion allowances, and their replacement by incentive grants of approximately equal value, but larger in size for firms with higher proportions of Canadian ownership.

The third case in table 13-7 involves all of the budget provisions of the second case plus two responses announced by the premier of Alberta on

October 30, 1980: a progressive cutback totaling 15 percent in the production of conventional crude oil; and indefinite postponement of any further synthetic-oil plants. The fourth case involves a compromise between federal and provincial interests. It increases the wellhead prices of crude oil by $2 per barrel per year in 1981, 1982, and 1983. In return for these higher prices, Alberta is assumed to restore the production levels for conventional oil, and to permit construction of a continuing sequence of new oil-sands plants.

Electricity, Nuclear Energy, and Coal

The electricity situation is far removed from that of oil and gas. In each province except Alberta there is a dominant electric utility that generates, transports, and distributes the bulk of the electricity sold in that province. Most of the Crown corporations and the few remaining privately-owned utilities are subject to some form of provincial regulatory control. There are some interprovincial and international grid connections, and hence some potential for NEB control of exports, but the trade flows tend to be incidental to the planning and operation of the utilities.

Nuclear energy, by contrast, is dominated by the federal government, which has played key roles in the international marketing of uranium; the design, construction, and sale of the CANDU reactor and the associated heavy-water plants; and the regulation of trade and safety of nuclear materials. The provincial utilities, on the other hand, are the only potential Canadian users of nuclear generators.

Coal is used by provincial utilities and steel plants (chiefly in Ontario) for thermal-power generation and steel making, respectively; it is produced in the same provinces that produce the oil and gas, with the important addition of the maritimes. The eastern coal production, like that in many other countries, is a subsidized declining industry. The western coal production growth has been chiefly of metallurgical coal for export. The coal for Ontario is mainly imported. Though there is clearly a substantial prospect for increasing coal exports, and for some increasing use of coal for conventional thermal generation of electricity, the longer-term energy use of coal may well be as a source of petrochemical feedstock, of liquid fuels, and of synthetic gas. In the meantime, there appears little prospect that coal will play a significant role in the design and implementation of federal energy policies. A national coal policy has been promised, but it is uncertain what the federal government can contribute beyond some support for improved transporatation facilities.

Natural Gas: Supply, Demand, and Pipelines

Supply Optimism in the 1960s

Natural-gas policy in Canada in the 1970s was dominated by the issue of Arctic pipelines. At the beginning of the decade, when the outlook for supply and demand was based on the 1969 NEB report which projected that domestic and expanded export sales could be met easily until 1990 and beyond from nonfrontier sources, there was no obvious link between domestic demand and northern pipelines. In 1970, the federal government established a *corridor concept* for northern pipelines, envisaging one trunk line for crude oil and another for natural gas. The federal government took an active supportive role in relation to the early proposals, which envisaged transport of Alaskan oil and gas, as well as the tapping of Canadian deposits for export.

Uncertainty in the Early 1970s

By 1972-1973, the situation had changed in several respects. The oil pipeline had been preempted by the Trans-Alaska pipeline, and there was a single natural-gas-pipeline proposal being advanced. The slow pace of nonfrontier discoveries and the fast growth of demand in the early 1970s combined to stop the approval of new exports. The 1973 report by the Department of Energy Mines and Resources (EMR) was already setting limits to the expected nonfrontier supplies before the world events between October 1973 and March 1974 combined to raise the market values of all energy sources. The NEB's 1970 approval of a 50-percent expansion of natural-gas export commitments now drew extensive criticism for excessive size, excessive contract length, and low prices. In mid-1974, the Alberta Gas Trunk Line left the Canadian Arctic Gas Pipeline Limited (CAGPL) consortium and proposed the Maple Leaf Project in collaboration with Westcoast Transmission Limited, through a jointly-owned subsidiary called Foothills Pipelines. The Maple-Leaf sponsors proposed to serve Canadian interests better by using existing pipelines in the south and a 42-inch pipeline in the north, carrying only Canadian gas by the Mackenzie Valley route to serve only Canadian markets.

Mid-1970s Pessimism and the Push to the Arctic

In this early-1970s climate of uncertainty, the National Energy Board started hearings in the fall of 1974 to determine the supply of and requirements for Canadian natural gas. The key participants were the two Mackenzie Valley pipeline groups, both committed to showing Canadian need for frontier gas to buttress their pipeline applications. Other key participants were the major producing companies, still in the midst of the federal-provincial-industry struggle for shares of the higher potential oil and gas revenues. The producing firms were anxious to show that expensive new sources of energy would have to be tapped, thus requiring higher prices and lower traxes.

Although the author argued before the NEB that even the industry data on demand and deliverability did not support the industry conclusion of early Canadian need for Arctic gas, the NEB came out strongly in support of the view that Canadian demand was already exceeding Canadian supply less existing export commitments, and would continue to do so until Arctic supplies were tapped. The NEB's conclusion that there would be continuing excess Canadian demand required an impossibly high demand forecast, unprofitably slow extraction rates, and the assumption of several unnecessary restrictions on the rate at which reserves could be connected and used.

Despite its very weak analytic underpinnings, the 1975 NEB report became the received government, industry, and public opinion about the supply and demand situation. It was the last official federal energy-policy report that did not take some explicit account of the effect of higher energy prices on energy demand. In the meantime, an econometric model of total primary-energy demand and energy shares was being developed within the Department of EMR.

The 1976 EMR report showed an apparent domestic need for frontier gas in 1985 of .2 tcf per year, only one-fifth as large as the NEB had been forecasting a year earlier. Although much more reasonable on the demand side, the EMR forecasts were still very weak, and much too low, on the supply side. They took inadequate account of the incentives for profitably faster production, and virtually ignored the inevitable explosion of exploratory drilling in the deeper basins of the Foothills region and elsewhere in response to the sixfold increase in natural-gas wellhead prices that had taken place between 1970 and 1976.

Alaskan Highway Route Chosen for Alaskan Gas

Things became more complicated in early 1977. Foothills Pipelines, in response to U.S. preferences, converted its Alaska Highway proposal to an Express Line for U.S. gas. Thereafter, as the Berger Report came out strongly against any Mackenzie Valley pipeline for ten years, and against either of the CAGPL routes across the Mackenzie Delta forever, on environmental grounds, Footshills Pipelines gradually backed away from its

Mackenzie Valley line. This left the main contest, in the final rounds of the NEB Northern Pipeline hearings, to be between the CAGPL Mackenzie Valley line and the Foothills Alaska Highway line. At this stage, Alberta Gas Trunk Line, the main Foothills sponsor, became much more receptive to the academic research submitted before the NEB showing that no frontier gas would be required for Canadian use until about 1990 or beyond, and that earlier development of the frontier gas would cause nonfrontier low-cost sources to be denied market access, thus sharply lowering the returns to nonfrontier producers and to the country as a whole.

Alberta Gas Trunk Line, based in Alberta, must also have become aware of the increasing fears of the successful exploring companies that they would be unable to sell their gas. Those successful companies, which did not participate directly in the NEB Northern Pipeline hearings, were nevertheless making themselves well known among policymakers. They feared that a NEB decision in favor of CAGPL, and based on Canadian need for frontier gas, would not only deny Canadian markets to the nonfrontier producers (since it was expected that the frontier gas would have preferred market access, just as does the Syncrude oil-sands production), but would make it impossible for export markets to be opened up again without the NEB, the government, and the industry losing all semblance of public credibility. While these arguments may have had impact in some circles, they did not visibly affect the NEB Northern Piplelines Report of July 1977. The demand estimates in this report were now substantially lower than in the 1975 NEB report and were more directly based on a set of price-responsive demand equations. The conclusion that some frontier gas would be required by 1983 to 1985 was only obtained by inconceivable restrictions on the supply estimates. The inconceivability comes about because the NEB assumed a set of production weights on existing reserves and production from new discoveries that would only produce 90 percent of the marketable reserves, even with an infinitely long production life. This mistake, which contradicts the meaning of the reserves figures (which are supposed to represent the sum of expected future production within the economic lifetime of the reserves), had the effect of decreasing the current and future stock of reserves by 10 percent. This, when coupled with an excessively conservative estimate of new discoveries, enabled the 1977 NEB report to show a possible Canadian need for frontier gas by the mid-1980s.

The NEB's 1977 supply-and-demand analysis, when combined with their acceptance of Berger-style reasoning on the Mackenzie Delta route, gave rise to the recommendation that the Alaska Highway pipeline be accepted, but only on the condition that Foothills Pipelines make an application by July 1, 1979 to build a spur pipeline down the Dempster Highway to connect with Canadian gas in the Mackenzie Delta. This was the line that eventually became the basis of the Canadian-U.S. pipeline agreement.

Late 1970s Pressure from Excess Gas Supply

As the prospect of imminent Canadian need for frontier gas (either from the Dempster spur to the Mackenzie Delta or from the Polar Gas and Arctic LNG projects) began to look more implausible, all of the producers achieved unanimity once more, this time in arguing for an expansion of exports as a necessary impetus to further exploration. It was proposed that expanded exports would be transported by pre-building the southern sections of the Alaska Highway pipeline, or, according to a competing proposal, by expanding TransCanada Pipeline's facilities. Also under consideration, although with less unanimity of producer support, was the eastward extension of the pipeline transmission system either to Quebec City (a TransCanada proposal) or to Quebec City and the Maritimes (a proposal of Alberta Gas Trunk Line). A necessary feature of either of these proposals (especially the latter) would be either a large federal-government subsidy or a corresponding drop in producer prices to market the small quantitites of expensively transported gas to areas better served by coal or imported residual fuel oil. A selective price cut only on gas marketed east of Montreal might represent a profitable source of extra sales for producers. From the federal government's point of view, the further eastward penetration of Alberta natural gas was regarded as a federal asset in the Quebec-independence debate, as well as further implementation of the announced goal of energy self-reliance. The persisting chance of substantial natural-gas discoveries off the coast of Nova Scotia presents at least the possibility of an eventual pipeline starting in the Maritimes and shipping gas westward to central Canada.

Additional Gas-Export Approvals

In the 1979 natural-gas-export hearings, the NEB followed the general industry view by increasing its potential-supply estimates on the basis of recently publicized discoveries in the Elmworth field and elsewhere in the so-called Deep Basin of the British Columbia-Alberta northern-foothills region. Treating those discoveries (which were well known but not widely publicized until mid-1978) as geological anomalies rather than as areas for the expected development of higher-cost drilling allowed the NEB to change its stance without rejecting its own earlier analysis.

In December 1979 the NEB recommended an additional 3.75 tcf of new gas exports, and this recommendation was quickly accepted by the federal government. By approving that amount of additional exports between 1980 and 1987, the NEB filled all producer requests for new exports prior to 1985, with a phasing out of the additional exports between 1985 and 1987. It is likely that subsequent hearings will be held in the early 1980s to expand and extend these permits until the originally requested 9 tcf (over a fifteen-year period) is reached or exceeded.

Federal Oil Policies

The supply of crude oil is the vulnerable part of the Canadian energy-supply situation. In 1979, Canada had net crude-oil imports of 215 thousand barrels per day, about 12 percent of total Canadian oil consumption. In the National Energy Program of October 1980, the federal government proposed supply incentives for tertiary recovery oil (with a reference price of $30 per barrel, adjusted for inflation), for oil-sands oil (with a reference price of $38 per barrel), and for oil drilling by Canadians on the Canada Lands (with incentive grants and tax deductions covering more than 90 percent of exploration costs). But the "centre-piece of the National Energy Program is a drive to reduce oil consumption, through conservation efforts and the use of more plentiful fuels in place of oil."[3] By a mixture of conversion incentives and subsidies for insulation, natural-gas pipelines, and renewable energy sources, the federal government proposes to reduce the share of crude oil in primary-energy demand from its current 43 percent to 27 percent in 1990. These policies are estimated by the federal government to cost $11.6 billion in the period 1980-1983, of which $2.5 billion is for exploration incentives, for upgrading of heavy oil, and for nonconventional oil. Over the same period, the federal government proposes to raise $5.2 billion from the new petroleum-and-gas-revenues tax and $7.3 billion from the new natural-gas and liquids tax. These new taxes, coupled with the relatively low wellhead prices for conventional oil and natural gas, have been the source of the main opposition to the program by the oil industry and by the governments of the producing provinces. Two other points of industry objective to the program have been:

1. Petro-Canada's right to 'back in' to a 25-percent interest in oil and gas production in the Canada Lands (in the Arctic and offshore areas); and
2. The plan to increase Canadian ownership in the oil and gas industry from about 33 percent now to 50 percent at the end of the 1980s, by various measures including purchase by Petro-Canada of several of the large oil and gas firms, to be financed by special charges on oil and gas consumption in Canada.

Electricity

Although capital expenditures for electricity generation are forecast by the federal government to account for over 60 percent of their 1977 forecast of

1976-1990 energy-capital expenditures, totaling over 180 billion 1975 dollars, the federal government has little scope for affecting the main electric-power decisions in Canada. The key issues relate to the amount and type of new generating capacity to be provided, and the rate structures that are to distribute the costs of electricity to the users. In contrast to oil and gas, where primary concern focuses on the distribution of economic profits among regions of the country, and between Canadians and the mainly foreign-owned oil companies, the distribution issues in electricity arise mainly between classes of users within each province. The economic costs of inefficient provincial electricity supply and pricing policies are thus borne almost entirely by residents of each province.

The federal government has two indirect roles in electric-power planning, the first being the integrated forecasting of all energy supplies and demands, and the second the production and regulation of nuclear reactors.

Inflated Estimates of Electricity-Demand Growth

In the function of integrated supply-and-demand forecasting, the federal government has no doubt found itself in an embarrasing position. On the one hand, the estimates of capital expenditures and system expansion are based on the plans of the major provincially based electricity-supply utilities. On the other hand, the demand forecasts used by the utilities in their 1975 expansion plans surveyed by EMR are so high as to be inconsistent with any of EMR's equation-based forecasts. Part of the problem is that the provincial utilities tend to have considerable freedom to expand their systems according to their own estimates of demand and then to set pricing schedules that recover their total costs and encourage the expansion of demand to meet the level of prebuilt capacity. The existence of excess capacity thus tends both to perpetuate itself, by delaying the rate-structure reforms required to match future marginal costs and revenues. It can be argued that the combination of high general inflation rates and high interest rates is putting enough financial pressure on the provincial utilities that they are likely on their own initiative to eventually wish to use rate-structure reform to cut the rate of growth of their expenditures to recover a higher degree of financial self-sufficiency. In the meantime, however, there is the prospect of an extended period of excess electricity supply, with capital and operating costs at unnecessarily high levels and with rate-structure reforms delayed as a way of absorbing the excess capacity.

Nuclear Energy

The federal government's second main role in electricity planning is as developer, regulator, and seller of nuclear generation facilities. This involves several conflicts of interest, as reduced electricity demand means fewer sales of CANDU reactors. Similarly, tighter safeguards on the domestic and foreign use and control of uranium and radioactive wastes mean fewer sales of CANDU reactors. The Canadian federal government, its Crown corporations, and its regulatory agencies probably have a deeper involvement in and commitment to nuclear power than does the national government in any other country. As has been shown in hearings before the Ontario Royal Commission on Electric Power Planning, there is considerable mistrust of the nuclear industry, and little public faith in the federal government's ability to deal objectively with the issues. On the touchy question of nuclear wastes, the federal government has recently commissioned a report from outside experts, and the government of Saskatchewan has recently had a Royal Commission on the safety of potential uranium-mining projects in the province. Both reports concluded that the risks were manageable, although antinuclear groups have presented counterevidence on the nuclear waste issue to the Resources Committee of the House of Commons.

The economic analysis of nuclear power is bedeviled by the fact that so much of the past and current development is directly or indirectly financed by government in ways that make comparable cost-estimating difficult or impossible to achieve. For example, 70.6 percent of federal direct-research expenditure on energy (excluding Atomic Energy Control Board grants) in 1972-1973 was for nuclear energy. This issue, as well as the related issues of government involvement in exports of uranium and reactors, must lie beyond the scope of this survey. What is apparent, however, is that even the direct costs of nuclear power generation are very high, and that reductions in the expected growth of electricity demand will be matched by very substantial reductions in the planned program of nuclear power development.

Conclusions

Unsupported Swings in Government and Industry Forecasts

During the 1970s there were unjustified swings in government and energy-industry forecasts of future energy supplies and demands, and hence swings in the appropriate policies influencing taxation, government support of energy-supply projects, and energy trade. Although the rapidity of the changes in opinion, and the resulting loss of credibility, have led to substantial attempts to improve the quality of official forecasting of energy demand, the analysis of energy supplies and costs is still primitive, and tends to be excessively dependent on the currently perceived state of confidence within the energy industry. This is particularly so in the case of crude oil and natural gas, where the price dependence of production flows, recovery factors, drilling effort, and ultimate reserves is not well understood.

Assuring Supply for Untested Demands

Energy policies in Canada at the federal and provincial levels have been excessively concerned, especially before October 1980, with increasing energy supplies, without ensuring that the demands (for appropriate uses) were likely to be large enough to justify the new high-cost projects. The federal government's attraction to large new projects was evidenced by temporary or sustained support of Syncrude, Arctic pipelines, Arctic and offshore oil and gas exploration, eastward extensions of natural-gas and oil pipelines, and nuclear power. The provinces, through their ownership or regulation of electrical utilities, have been mainly responsible for overbuilding electricity supply.

As for the design of pricing policies to ensure an efficient pattern and suitable level of demand, no Canadian government gets high marks. The federal government's idea of moving oil and gas prices toward world levels in a series of preannounced steps offered the prospect of reducing the short-term distribution effects and adjustment costs, while encouraging energy users to think of world prices when designing their new plants and homes. However, the plan has been marred in practice by continuing indecision about exactly where the prices are going and when they will get there. The general uncertainty about energy price levels is compounded (with some reason) by uncertainty about the world price of crude oil, and about the external value of the Canadian dollar. Less justifiably, federal and provincial governments are both inclined to use selective cuts in energy prices, or energy subsidies of some sort, to implement their policies of industrial strategy (for instance petrochemical developments) or energy substitution (such as subsidies to extend natural-gas use to the Maritime provinces). On top of this, none of the provinces has yet reformed the rate structures of its energy utilities to ensure that each user pays prices that reflect the costs of new supply.

The Regulatory Lag

The often-decried lags in getting approval for large new energy projects have often performed a valuable function in reducing the extent to which supply is likely to overshoot demand in the 1980s. If the Berger hearings had not been so drawn out, and if the Berger Report had not had such a wide public impact, the federal government would have accepted one of the

Mackenzie Valley pipeline proposals on terms that would have involved expensive subsidies, from taxpayers and energy consumers, for natural gas that would have been much more expensive than the alternative sources of supply. By the same token, the much shorter Thompson West Coast Oil Ports Enquiry was partly responsible for decisions being deferred long enough that it could be fairly easily demonstrated that there would be no short-term Canadian need for such a facility.

However, on the negative side, some of the same inertia has acted to slow down moves toward energy conservation, toward revision of utility rate structures, and toward the development of potentially cheaper alternative energy sources.

Conflicts of Interest

Both levels of government have equipped themselves with so many conflicting roles in energy management that they have lost their ability to set guidelines and make regulatory decisions without facing severe conflicts of interest. Examples from the survey include Panarctic Oil's drilling in the Arctic under federal government environmental controls; Petro-Canada being the largest shareholder in Westcoast Transmission, which is not only a federally regulated pipeline but also a partner in the Alaska Highway pipeline; the Polar Gas project sponsored by a Crown-controlled corporation; the federal government selling uranium and reactors while also regulating safety in the handling and disposal of radioactive materials; petrochemical developments under both federal and provincial government sponsorship; and provincial regulatory commissions controlling provincially owned utilities, both being subject to direct or indirect cabinet control.

Energy Self-Reliance as a Policy Goal

The policy goal of Canadian energy self-reliance has been only vaguely defined, yet has been widely accepted by parties with sharply conflicting interests. If it is taken to mean only that policies should, other things being equal, facilitate the meeting of legitimate energy demands in Canada in the least costly way, bearing in mind the costs imposed if offshore sources or expected new technologies should be unavailable, then it is only a rather imprecise way of stating that energy resources should be developed and used in the most efficient manner. If, in contrast, it means that Canadian energy needs should be met to the maximum extent possible from Canadian sources and that Canadian energy resources should be devoted to meeting

Canadian needs, then it involves considerable potential for inefficient use of resources. To strive for self-sufficiency in every key sector means giving up the possible gains from trade with other countries that have different tastes and resources.

Under either of the above interpretations, the goal of national energy self-reliance is vague about the international, interregional, intertemporal, and interpersonal distribution of the costs and benefits of energy production and use. Yet it nevertheless has a national flavor. Thus, it provides a veneer of public interest while providing plenty of scope for private interests to interpret the policy in their favor. Therein lies the key to its apparently widespread acceptability. For example, the Canadian Petroleum Association, the chief representative of the major international oil companies, when speaking in 1975, used self-sufficiency as grounds for higher prices, lower taxes, and lower royalties, all with the idea of increasing supply. With that goal accomplished, the industry focus then changed, and the international companies tended to argue that more exports are now required to make markets for current surpluses, in order to encourage and finance further exploration that will serve to provide furture Canadian needs.

Canadian users of energy, to the extent that they are organized at all, tend to agree with Canadian nationalist groups that energy exports and domestic prices should both be kept low, with the twin aims of reducing costs to Canadian users and decreasing the share of economic rents from energy resources that fall into the hands of international oil companies or into the provincial coffers of Alberta.

The Maritime and central-Canadian provincial governments support the above position to the extent that they are not able to get a direct or indirect share in the Alberta revenues. Alberta applies a similar policy within Alberta and accepts certain aspects of its applications in the rest of Canada as part of an informal interregional aid program, making varied use of the political capital thereby obtained.

The vagueness of the self-reliance goal may have helped it to gain at least superficial acceptance; but this means that the announced goal of energy policy has few particular implications for energy policy and is difficult to relate to more fundamental goals of efficient energy use and equitable distribution of costs and benefits. The Iranian political crisis of 1979 and the associated doubling of the world price of crude oil have done much to improve the economic attractiveness of many Canadian energy-supply projects. In addition, the increasing unreliability of world oil supply makes a policy of Canadian self-sufficiency in crude oil less costly and more feasible than before. As a consequence of these higher prices and less certain world supplies, Canada seemed likely to remain a net exporter of energy, and to be only a modest importer of crude oil and petroleum products.

However, the sharp increase in world oil prices also shattered the fragile agreement between Alberta and the federal government about the appropriate sharing of the economic surpluses created by the higher world prices. After defeating the federal Conservative government over the energy prices and energy excise taxes proposed in the budget of December 1979, the Liberal government was elected in February 1980 with a promise to keep energy prices to users lower than was proposed by the Conservative budget. The federal National Energy Program of October 1980 attempted to redeem this pledge while at the same time increasing the federal government's share of energy revenues. This meant sharp reductions in the current and anticipated revenues of the producing firms and the governments of the producing provinces, especially during the first half of the 1980s. The unilateral setting of wellhead prices by the federal government, and the imposition of commodity and wellhead taxes on gas and crude oil, prompted the Alberta government to propose a phased 15-percent reduction in conventional crude-oil production in 1981, and to postpone indefinitely the approval of any new synthetic-oil projects in Alberta. In addition, court cases were instituted by the producing provinces to test the constitutionality of the natural-gas tax in those circumstances when it could be held to amount to taxation of one government by another.

In early 1981 this stalemate is still unresolved. In the short term, at least, the effect is to lessen the efficiency of energy use and the level of Canadian self-sufficiency in crude oil. This in turn will lead to marginal increases in the tightness of the world oil market.

Epilogue

In September and October 1981, energy pricing and taxation agreements were negotiated between the federal government and the governments of each of the three main producing provinces. These far-reaching agreements reflect the increasing political pressures on both levels of government to end the energy stalemate that had existed ever since the introduction of the National Energy Program (NEP) in October 1980. Many major energy users were advocating a resolution even if it meant higher prices. The producing firms were feeling the effect of the Alberta production cutbacks, as well as the drop in exploration activity, and were putting pressure on both levels of government to reach an agreement that would create a more secure climate for longer term planning. These mounting pressures undoubtedly made it easier for the two levels of government to reach an agreement with much higher prices than the federal government could have accepted earlier, and much higher federal taxes than the Alberta government could have found acceptable.

Several main points have emerged from analysis of the new energy agreements:[4]

1. The new agreements commit both levels of government to the end of 1986, and thus provide a much more predictable structure than in any recent period.

2. Prices for oil and natural gas users, and for new supplies of conventional and nonconventional oil, are now linked very closely to the levels of world prices, unless world prices should rise faster than 10 percent per year.
3. The substantial increases in the prices to be paid by energy users and to suppliers of new oil will reduce demand, increase new discoveries, and cut net imports substantially.
4. The macroeconomic effects of the agreements, when compared with the previous stalemate, will lead to higher levels of investment, output, and employment. However, as an offset, the consumer price index will also be higher.
5. Under the agreements, the federal and provincial governments are, to an almost equal extent, much better off than under the stalemate (or the NEP). The producing industry is also better off, although to a lesser extent, and energy users are substantially worse off. The average Albertan is much better off with the agreements, while the average Canadian outside Alberta is unaffected.
6. The federal share of economic rents from oil and natural gas will increase substantially during the life of the agreements, offset by reduction in the share going to energy users.
7. In terms of supply effects, the most important difference between the energy agreements and the NEP comes from offering the world oil prices for new discoveries of conventional oil. This gives after-tax wellhead prices and discoveries that are almost twice as high as under the NEP, especially during the early years of the agreements.
8. Actual revenues are likely to be less than those forecast in the agreements, especially for the producing industry.
9. World oil prices are not rising as fast (as least until the end of 1982) as forecast in the agreements. This provides an additional reason for revenues to be lower than forecast.
10. Overall, the agreements provide a more efficient and stable structure of energy prices and taxes than seemed likely to emerge when the conflict began.

Notes

1. This chapter is an updated and much abridged version of John F. Helliwell, "Canadian Energy Policy," *Annual Review of Energy,* 4 (1979), pp. 175-229 (reproduced by permission of Annual Reviews Inc., Palo Alto, California). Complete footnotes, references, acknowledgments, and additional tables are to be found in the full version. Much of the same material also appears in *The Canadian Economy: Problems and Options,* R.C. Bellan and H. Pope, eds. (Toronto: McGraw-Hill Ryerson, 1981).

2. *The National Energy Program* (Ottawa: Department of Energy Mines and Resources, 1980), p. 52.

3. Ibid., p. 99.

4. This summary is drawn from John F. Helliwell and Robert N. McRae, "Resolving the Energy Conflict: From the National Energy Program to the Energy Agreements," *Canadian Public Policy* 8:1 (1982).

14 Japanese Energy Policy

Ronald A. Morse

For nearly a decade the politics of Middle Eastern oil has had a major impact on Japanese domestic economic planning and global diplomatic strategy. The first oil crisis accelerated the pace of Japan's bilateral diplomatic ties with Middle Eastern nations and initiated a series of generally unsuccessful domestic economic policies to cope with higher energy prices. Following the Iranian revolution in 1979, Japan managed its economic programs more effectively and, though not abandoning its pro-Arab posture, became increasingly aware of its multilateral responsibilities in the international energy market. For the 1980s, Japan will continue this strategy while emphasizing energy efficiency in its industrial production and promoting commercially attractive alternative-energy technologies.

Japan's comparatively successful adjustment to higher oil prices has led many energy specialists to conclude that Japan has already turned an important corner in its long-term effort to reduce its energy demand. While Tokyo appears to have brought its demand for imported oil under control, its medium- to long-run pattern of energy supply and demand will remain based on oil. Because of its highly concentrated urban population centers, narrow coastal industrial zones, and strict environmental regulations, Japan cannot easily swing away from petroleum use to alternative or new energy sources.

As the world's second-largest importer of petroleum, Japan's policies will have a significant impact on future world-energy markets and prices. In 1980 Japan paid $53 billion for approximately 5 million barrels of oil a day (mbd), over 10 percent of free-world production. Nearly 80 percent of Japan's oil presently comes from the volatile Persian Gulf region.[1]

The Energy-Policy Process

The Japanese policy process is based upon a slow, mid-level-bureaucratic, group-consensus process that emphasizes continuity and the priority of maximizing Japanese economic interests. The Japanese apply a passive, adaptive process to new situations; a tactic that is pragmatic and not overly reactive. A close but informal consultative mechanism between industry and government that seeks to maximize market forces is also involved.

Table 14-1
Outlook for Primary-Energy Supply and Demand in Japan

	Fiscal Year									
	1979 (actual)		1985				1990			
			Forecast by the Government[a]		Forecast by the IEE[b]		Forecast by the Government[a]		Forecast by the IEE[b]	
Type of Energy	*Actual Number*	*Percentage Of Total*	*Actual Number*	*Percentage of Total*	*Actual Number*	*Percentage of Total*	*Actual Number*	*Percentage of Total*	*Actual Number*	*Percentage of Total*
Total primary-energy demand *(hundred thousands of kiloliters)*	444		482		582[c]		551		700	
Hydroelectric power (general) *(thousands of kilowatts)*	18,800		22,000		20,000		26,000		23,000	
		5.0		4.7		5.3		4.6		5.4
Hydroelectric power (pumping-up), *(thousands of kilowatts)*	9,500		19,500		17,000		27,000		22,000	
Geothermal heat *(thousands of kilowatts)*	160	0.1	1,000	0.4	350	0.0	3,500	1.2	1,000	0.1
Domestic pertroleum and natural gas *(thousands of kilowatts)*	3,250	0.7	8,000	1.4	5,380	1.1	9,500	1.4	5,880	1.1
Domestic coal *(thousands of tons)*	18,000	2.9	20,000	2.5	19,030	2.6	20,000	2.0	19,030	2.3
Nuclear power *(thousands of kilowatts)*	15,100	4.2	30,000	6.7	23,000	6.7	53,000	10.9	36,000	9.3
Liquid natural gas *(thousands of tons)*	14,900	4.7	29,000	7.2	26,950	8.0	45,000	9.0	36,150	9.4
Imported coal *(thousands of tons)*	59,400	10.9	101,000	13.6	89,610	14.1	143,500	15.6	119,930	16.3
Steam coal *(thousands of tons)*	(1,680)		(22,000)		(19,230)		(53,500)		(45,760)	
New forms of energy *(thousands of kiloliters)*	500	0.1	5,200	0.9	1,290	0.3	38,500	5.5	4,230	0.8
Subtotal *(hundred thousands of kiloliters)*	127	28.6	216	37.1	185	38.1	350	50.0	245	44.5

Imported oil *(hundred thousands of kiloliters)*	317	71.4	366	62.9	299	61.9	(350) 366	50.0	305	55.5
Liquid petroleum gas *(thousands of tons)*	(9,700)	(12.5)	(20,000)		(12,640)		(26,000)		(14,570)	
Total energy supply *(hundred thousands of kiloliters)*	444	100.0	582	100.0	482	100.0	(700) 716	100.0	551	100.0
Balance *(thousands of kiloliters)*							16,000			

Source: *The Energy Supply and Demand Structure Outlook in Japan*, (Tokyo: Long-Term Credit Bank of Japan, 1981), p. 14. Reprinted with permission.

[a]The economic growth rates, 1985/1976 and 1990/1985, are placed at 5.5 and 5 percent respectively by the IEE. These estimates are a little more conservative than those of the government. The GNP elasticity of energy demand is also placed at a conservative value of 0.62 (1985/1973). Japan's GNP elasticity of energy demand agreed on at the Venice Summit Meeting held in June 1980 is 0.6 or less.

[b]IEE is the Institute of Energy Economics. The forecasts by the IEE were announced in February 1981, and those of the government were announced by the Advisory Committee for Energy in August 1979.

[c]After an energy-saving drive.

Government agencies are the main actors in energy policy. Continuity in government, especially in the bureaucracy, helps explain why Japan has handled the second oil crisis much better than it handled the 1973 crisis. Five prime ministers, all from the conservative ruling Liberal Democratic Party, have directed Japan's energy policy since 1973. The policy planners have been the same people over the years. By the time a policy paper reaches the cabinet-ministerial-level energy-advisory councils and other decision makers, a government- and industry-wide deliberative and consultative process has already taken place that virtually ensures the recommendation of the policy proposals. The key administrative organizations responsible for this process are the Science and Technology Agency, responsible for nuclear research and development, and the Ministry of International Trade and Industry (MITI), which has central responsibility for most areas of energy policy. The Agency for Natural Resources and Energy within MITI was created just before the 1973 oil crisis and remains the focus of Japan's energy policies.

In the consultative process the Ministry of Education (university and research funding), the Environment Agency (pollution policy and regulation), the Ministry of Finance (budget support for national energy policies), and several other economic planning bodies have a significant input. The Ministry of Finance has the critical say on funding for energy R&D and budgetary support for energy programs. The Japan Atomic Energy Commission decides on matters related to research, development, and utilization of nuclear energy, including regulatory and licensing matters.

The Japanese parliament has special committees on energy policy in both upper and lower houses. Apart from the government structure, the ruling Liberal Democratic Party's Political Affairs Research Committee makes its own policies on energy and R&D matters, and reviews government energy-legislative bills and programs. Opposition political parties have a separate agenda but are essentially excluded from the energy-policy arena.

The second major group is a loose alliance of business and industry leaders. Private industry has an input into national energy policy through its participation in government advisory bodies and through industrial federations or specific industrial lobbying groups like the Japan Coal Association or the Electric Power Industrial Association. The large Japanese trading companies, often ignored in this process, actually handle 60 percent of Japan's energy imports and exports. At present they handle nearly one-third of Japan's oil-market purchases. Their views are reflected through a wide range of industrial and corporate affiliations.

As new issues or initiatives arise (restructuring the oil industry, increased coal imports, energy assistance to developing nations, etc.), these groups or organizations make adjustments, reallocate resources, establish new committees, and call on different experts to examine and report as appropriate.

A cumbersome process, it has served Japan well during an era in which it has not been forced to take the lead on global political or strategic initiatives.

Finally, there are separate organizations, experts, citizens groups, and the media; all interested in energy issues. Environmental protest groups and local citizen associations opposed to nuclear and other kinds of energy facilities have been very effective in delaying or stopping a number of projects. Though these groups presently enjoy limited political backing from progressive political parties, they are still important and must be taken into consideration in the government planning process.

The First Energy Crisis

After World War II, Japan pushed ahead with an industrial strategy for economic growth that required high natural-resource imports and demanded access to foreign economies ready to purchase manufactured goods. In the process of building this economy, Japan, like many of the industrial nations, became heavily dependent on imported cheap petroleum as its primary energy source.

Before the 1973-1974 energy crisis the Japanese had given some thought to this change in its energy-supply profile but only in a passing way; Japan was satisfied to rely on the major international oil companies for its oil and stay out of Middle Eastern affairs. Japan established diplomatic relations with Israel in 1952. When the Suez Canal was closed in the crisis of 1956-1957 and the Iraqi pipeline cut, Japan experienced a dramatic rise in tanker rates because of longer shipping routes. Again, during the Middle Eastern war in 1967, although Japan was not the target of supply cuts, it did experience another increase in shipping charges. To the Japanese, the Middle East was remote and of little significance except as a supplier of petroleum.

The energy crisis of October 1973 motivated the Japanese to take a careful look at their interests in the Middle East. The domestic Japanese response to the crisis and daily commodity hoarding initiated by a fear of a continued oil disruption were things the Japanese had not experienced since World War II. The Japanese rushed to placate the Arab oil-producing nations. Japan followed the European community in its redefinition of "resource diplomacy."[2]

The decline in the Japanese oil supply in 1973 was modest: October, 5 percent; November, 9 percent; December, .1 percent; January, 13 percent; and February, 14 percent. Petroleum-product stocks dropped from thirty-two to twenty-five days and the crude supply dropped from twenty-six to twenty-one days—forty-six days' supply being only one day more than the amount required to maintain forty-five days of running stocks. The crisis, however, did have a major political impact, and produced the management legislation, regulations, and price guidelines that now provide the foundation for Japan's emergency preparedness strategy for possible oil disruptions.

Economic policies were mismanaged following the first oil crisis and pushed Japan into a recession. As one writer put it, "In short, the Japanese economy weathered the oil supply and pricing crisis by ruthlessly damping inflation, prolonging the recession, and achieving economic equilibrium on a smaller scale of industrial and import activities. Beginning in 1975, demand-restraint policies were gradually relaxed, and the underproducing economy began a gentle ascent toward recovery."[3]

In early 1973 Japanese officials made it clear that politically they desired closer bilateral ties with the oil-producing nations. They saw, as has continued to be the case, little value in cooperation among the major oil-consuming nations. "On the diplomatic front, the Tanaka cabinet's response was to tilt Japan's neutrality closer to Arab views in the hope that this would induce a lifting of the oil-supply restrictions on Japan. This new posture was announced publicly on November 22, 1973 by Chief Cabinet Secretary Susumu Nikaido in a statement to the press. The statement set forth Japan's insistence that Israel's forces be totally withdrawn from all the territories occupied in the 1967 war; that the just rights of the Palestinians, based on the U.N. Charter, should be recognized and respected (Resolution 242 had mentioned only a 'fair solution of the refugee question'); and, in a concluding passage, that the government of Japan will continue to observe the situation in the Middle East with grave concern and, depending on future developments, may have to reconsider its policy toward Israel."[4]

In December 1973 Deputy Prime Minister Takeo Miki and the international economist and later foreign minister Saburo Okita were dispatched to the Middle East on a twenty-day fact-finding mission. At the Foreign Correspondents' Press Club speech following the trip, Okita outlined what subsequently became Japan's policy for the 1970s. He said that there was a need for more direct contact with the oil-producing nations, that Japan had to gradually dissociate its policy from the pro-Israel position of the United States, and because of the high risks for private investment in the region, the Japanese government would have to assume a major role in the future bilateral dealings.

Sunao Sonoda, who has served as Foreign Minister twice, first brought high-level attention to the Middle East through his trip there in January 1978. He also paved the way for the visit by Prime Minister Fukuda in September 1978, the first visit to the Middle East by an incumbent Japanese prime minister.

The Second International Oil Crisis

Japan's more successful strategy for handling the oil crisis of 1978-1979 cannot be appreciated without considering the background circumstances and the special characteristics of Japan's economy. Japan benefits from a unique set of economic and political conditions: it is an attractive OPEC investment, Japan has a low expenditure (1 percent of GNP) for national defense, and it has been fortunate that it can transfer its more expensive energy costs by boosting exports. Like other countries, it borrows from the oil-surplus countries through capital markets. Whether Japan can continue to enjoy these benefits without creating political problems remains to be seen.[5]

The interlude between 1973 and 1979 was not wasted. The Japanese learned a great deal about energy-crisis economics following the 1973 shock. Japanese oil imports were essentially at the same level in 1973 (4.9 mbd) and 1978 (4.76 mbd). The Japanese response in 1979 took advantage of previous experience and changed conditions. In 1979:

Japan had nearly doubled its 1973 level of fifty-nine days of private and official emergency oil stocks, giving them more confidence;

The Japanese public was psychologically more aware of the problem of possible oil disruptions and the potential consequences;

The role of transactions by the international oil companies had declined to 45 percent in 1980 (it was 65.8 percent in 1978), and Japanese bilateral deals had increased;

Direct deals and government-to-government oil deals went from 5.4 percent in 1972 to 31.7 percent in 1979, serving as a buffer; and

Economics dictated a policy of tight money. Investment in energy-efficient facilities and technology continued with an emphasis on export competitiveness as domestic demand dropped.

The major supply cuts were to the Japanese oil companies not directly affiliated with, but dependent upon, supplies from the major international oil companies. The companies suffering cuts scrambled to the spot market to make up for shortfalls—this kind of disruption competition helped drive up the spot-market price for oil and strained relations among consuming nations.

The Japanese government responded to these cutbacks in two ways. MITI encouraged domestic oil companies and trading firms to secure crude oil directly from the oil-producing states. Secondly, MITI raised the price ceiling for spot-market purchases. Thus, in 1978 about 4 percent (.2 mbd) of Japan's petroleum was purchased on the spot market. This figure doubled by 1979 to 8 percent (.4 mbd) of Japan's imports and ranged from 6 to 10 percent in 1980 (.3-.5 mbd). The figures for spot-market purchases in November and December 1979 and January 1980 were between 14 and 18 percent.

In 1979, OPEC was also not unified and price increases were dramatic but more gradual. Other behavior (and here the Japanese are very much like the European oil-consuming nations) was similar to that in 1973: there was a scramble to make spot-market purchases in anticipation of shortfalls; multilateral cooperation was limited despite the IEA mechanisms; and the pro-Arab leanings continued despite the fact that Arab-Israeli issues were not directly related to the crisis. Building on the 1973 economic experience, in 1979 the government did not selectively try to employ price controls, wage increases (with labor union cooperation) were kept in line with low inflation rates (in 1974 wages went up 33 percent), and the national economy (the budget deficit) was managed well.[6]

Japan has continued to give priority to its close relations with the oil-producing nations. In July 1979, Masumi Esaki, the Minister of International Trade and Industry, returned from the Middle East convinced that Japan still required a more realistic assesssment of its regional interests. Responding to this on August 6, Foreign Minister Sonoda issued a Middle Eastern policy paper elaborating the Japanese government's increasingly sympathetic pro-Arab posture. He emphasized that Japan, free from the hindrances of a colonial power and with no ideological or political interest in the region, was free to develop a sound and independent relationship with the Middle East. Japan was never overly supportive of the Egypt-Israel treaty. In October 1981, Prime Minister Zenko Suzuki met with Palestine Liberation Organization (PLO) leader Yasir Arafat after nearly three years of cautious maneuvering to appear supportive of the PLO and still not rupture relations with the United States. Sonoda was foreign minister then too.

At the working level, the Japanese government and private groups have made every effort to become more directly involved in the Middle East. Since 1973, the Foreign Ministry has improved its staff of experts on the Middle East and the numbers of Arabists employed in the private sector has also greatly increased. Ambassadors to the Middle East have also assumed an active role in determining Japanese policy for the region. MITI now has an independent Middle East Division in its International Trade Bureau. Japanese parliamentarians, much like their congressional counterparts in the United States, have moved to establish their own independent expertise on the area. It is the Japan-Palestine Friendship League of parliamentarians that has become the major force in promoting semiofficial Japanese contacts with the PLO. All of this is just the beginning of an ever growing Japanese political and economic involvement in the Middle East.

The second oil crisis accelerated trends already under way in Japan: it boosted the shift to industrial energy efficiency and the implementation of a less-intensive energy structure. Japan now intends to continue its formula for success by diversifying its imports across fuel types and building stronger trade relations for its technology-based export products. Japan appears to have broken the link between oil consumption and the rise and fall of business activity. Since 1977 oil consumption has steadily declined, ushering in an era of possible zero energy growth.

Energy Emergency Preparedness

Since 1973, the Japanese have given increasing attention to the relationship between energy supply and security in an effort to reduce their economic vulnerability in an oil cutoff. This energy security is focused on short-term, sudden, or creeping types of disruptions that might take place in the next five to ten years, the period during which oil dependency will most likely remain constant. Japan's strategy for an energy crisis has been to minimize the domestic economic impact of a severe petroleum cutback.

Japan's crisis-management plan involves moderate government intervention in the oil market: the Japanese also have an elaborate set of regulations and established bureaucratic procedures to deal with future energy emergencies. As plans now stand, Japanese officials privately believe that existing multilateral arrangements (especially the IEA) will not work, that domestic crisis-management mechanisms will not be required for a short-term crisis (and would not prove effective in a sustained major crisis), and that Japan's only defense is to have adequate oil and other energy stockpiles to protect itself. While Tokyo finds the IEA useful as a source of oil-market information and as a focal point for energy-policy coordination, it does not have confidence in the emergency-preparedness system and has in fact entered into bilateral deals that are at odds with IEA policies. This reflects a more general view, held even outside Japan, that the world oil market alone cannot be relied on to distribute oil during a supply emergency.

The key supply-crisis regulation is the Petroleum Supply and Demand Adjustment Law, a law that was hastily compiled to cope with the 1973 crisis. Specific measures under the law include petroleum-supply targets, marketing and import plans submitted by refiners and importers, authority to modify such plans, authority to promulgate restriction on the use of petroleum, provision for mediation in disputes concerning oil allocation, and authority to impose allocation and rationing schemes. The law covers crude oil, gasoline, naptha, jet fuel, kerosene, gas, fuel oil, and liquefied petroleum gas (LPG).

A cabinet decision is required to impose supply quotas and to institute allocation and rationing systems. The preparations for handling rationing in the private and industrial sectors have been carried out by MITI under contract with private research institutions. The necessary rationing coupons have been printed, and they could be openly exchanged by those receiving them. It is hoped that this would prevent the development of a black market for fuel and other supplies.

There are a number of other issues that remain to be resolved. There is still no central-control command structure to manage the flow of imported oil, implement government policies, and distribute rationing coupons in the event of an emergency. How private and government oil-stockpile use would be coordinated remains unclear. The Japanese still have limited experience with fuel switching and power-supply sharing.

The Japanese government has been criticized for its failure to prepare the public adequately for the possibility of another energy crisis. MITI officials have been unresponsive on the issue and are reported to fear the possibility of having to resort to supply allocations or rationing. They have probably kept their emergency-preparedness plans secret for the wrong bureaucratic reasons; the real fear involves the Japanese national psychology. In Japan, where social interests receive a much lower priority than private or corporate interests in a conflict situation, an allocation or quota system may not work. The Japanese, who are used to preparing for national disasters if the source of the disaster is impersonal, may not respond in a cooperative way to an energy shortage in which it is possible to identify the source of the problem. People perceive the energy situation as a zero-sum game: a gain by one person necessarily means a loss by someone else. There is also the fact that oil-disruption costs may be borne without regard to income classes or geographical needs.

Japan has been willing to pay for crude-oil and oil-product storage. Its governmental petroleum-stockpiling program effort is managed by the independent public corporation, the Japan National Oil Company (JNOC). In 1978, the overseas oil-development responsibilities of the Japan Petroleum Development Corporation (1967) were combined with the national stockpile effort and the new JNOC absorbed both responsibilities. The Japanese government plan to increase its oil stocks entered a new phase recently with the start of construction of a permanent 35-mbd oil-storage facility in northern Japan. Due to be completed in March 1983, the 70-percent-government-owned oil-storage project includes floating storage complexes in which Japan's shipbuilding technology will be employed. Until these facilities are completed, Japan will continue its policy of offshore-tanker stockpiling. Onshore permanent storage will cost about two-thirds the cost of VLCC (Very Large Crude Carrier) floating storage and is much cheaper over the long run. At the end of 1981 Japan had 123 days of oil in private and government storage.

The momentum for increased stockpiles has come from MITI and indirectly from JNOC. Japan has continued to make stockpile purchases even under tight oil-market conditions. Less concerned with price and its reliance on international disruption-sharing schemes (IEA), Japan has given first priority to reducing vulnerability through stockpiling. Coal and LPG stockpiling is also now under serious consideration.

Energy Strategy for the 1980s

Japan's energy strategy for the 1980s is to shift away from energy-intensive industries and to give a high priority to energy conservation. The present optimistic plan is to reduce energy demand by 30 percent by the year 2000. Thus far industry, which uses nearly 60 percent of Japan's energy, has played a major role in Japan's energy-conservation effort; it has a major stake in the nation's ability to develop energy-efficient production techniques. Investment in the research and development of energy-efficient equipment has taken a dramatic step forward since the 1979 oil shock. Energy consumption in the residential-commercial sector (19 percent) is expected to remain constant. Private-sector consumption of energy will continue to rise as lifestyles continue to improve.

Japan is in good condition; its energy consumption per capita is nearly one-third that of the United States, and lowest of the industrial nations. Economically it has also done well; in 1980, Japan had a 5 percent growth rate in GNP, inflation was between 5 and 6 percent, and unemployment at a low 2 percent. Politically, Japan seeks to avoid taking a position; comfortable with a more passive international role, it has preferred to emphasize its energy-demand-reduction efforts and its interest in energy research for the development of energy-saving materials and techniques. Japan, despite its pro-Arab leanings in 1973 and preference for bilateral relations, has been constrained in its Middle Eastern strategy because it cannot export arms to the oil-producers and has a close security relationship with the United States. Japan, unlike the European nations, does not have a regional security alliance to fall back on.

Energy Conservation

There is no doubt that investment in energy conservation and the substitution of capital for energy will assume increasing importance in Japan's economic struggle against higher energy costs. A great deal has already been accomplished. In 1979, the economy grew in real terms by 6.0 percent, while energy demand increased by only 3.3 percent. The government has implemented a variety of demand-restraint measures, including voluntary guidelines and investment incentives to utilize energy-efficient equipment and increase conservation. Industry in Japan today is making large investments in oil-saving equipment; the oil consumption per unit of industrial output has already decreased significantly.

As Japan improves its use of imported petroleum, for the medium term it will make increasing use of liquefied natural gas (LNG), coal, and nuclear energy as oil substitutes.

Liquefied Natural Gas

In Japan today, in contrast with the United States and Europe, gas, and particularly liquefied natural gas (LNG), is increasingly used for power generation. Strict pollution regulations and the uncertainty surrounding nuclear power generation have made natural gas in the form of LNG an attractive energy option.

In 1980 Japan imported 16.84 million metric tons of LNG to supply some 15 million customers, a significant increase over the previous year. LNG imports currently meet 4.5 percent of Japan's total energy-consumption needs and are projected to reach 7 percent (29 million metric tons) of energy consumption by 1985. The target for 1990 is 45 million metric tons. Japanese LNG imports account for approximately 51 percent of world LNG trade, about 1.5 percent of world natural-gas consumption (the United States accounted for 39 percent and the Soviet Union for 26 percent). Nearly 75 percent of Japan's LNG imports are used by the major urban electric-power companies. This is largely because, unlike other fossil fuels, LNG contains virtually no sulfur or nitrogen oxides, the two pollutants controlled by environmental standards. LNG, despite its volatility, is not considered a dangerous fuel in environmental terms by the Japanese public.

Japan's success in increasing its LNG use will depend upon the availability of semigovernmental loans for the construction of liquefaction bases, transportation and distribution systems, and LNG tankers. The Japanese government currently supports the LNG industry through free import duties, loans and/or investments for natural-gas exploration, and the construction of liquefaction plants in producing nations. There are also financial inducements for LNG ship construction and receiving terminals through the Japan EXIM Bank, the Japan Development Bank, and the Japan National Oil Corporation.

Nuclear Power

As of June 1, 1981, there are twenty-two operating nuclear power plants in Japan, with total output of 15,511 million kilowatts, generating more than

12 percent of Japan's total power capacity. Japan had eleven plants under construction in 1980. By 1995 the power percentage is projected to be increased to 27-28 percent, with nuclear power supplying 37-39 percent of the nation's total power supply (as forecast by the Nuclear Energy Division of the Overall Energy Council, and advisory organ to MITI). In addition to its nuclear capacity, Japan has an established nuclear industry and R&D capability. Japan is continuing its research and development on fuel enrichment and reprocessing with the ultimate objective of a complete domestic nuclear-fuel-cycle capability.

Progress in the development of nuclear technology in Japan, not surprisingly, has been the result of close government-industry collaboration in the areas of industrial structure, R&D, siting and licensing, and in various international activities ranging from uranium acquisition to nonproliferation.

The budgetary support for nuclear power has been generous; the 1980 nuclear budget was 24.6 percent above that of 1979, and in 1981 there will be a 9.6-percent increase. Expenditures for nuclear R&D are by far the single largest science-budget item—approximately 18 percent of the government expenditure on science and technology.

Tensions with the United States over nuclear reprocessing were alleviated in September 1981 when the United States agreed to lift almost all of its previous restrictions on the reprocessing of spent nuclear fuel. The Reagan Administration agreed to lift the cap on how much spent fuel can be reprocessed at the Tokaimura plant, which went into commercial production in January 1981. Under the new arrangement, the plant will be allowed to operate until the end of 1984, by which time a permanent solution to the problem of the continuing functioning of the facility will be reached. The United States also agreed to lift all restrictions on the construction of a large-scale, privately owned spent-fuel reprocessing plant in Japan, although it will have a say in the facility's eventual operations.

Nevertheless, the siting and safety constraints on nuclear power development remain significant. The discovery of radioactive leakage from the Tsuruga nuclear power plant of the Japan Nuclear Power Company in April 1981 exacerbated public fears about nuclear energy and quickly took on aspects of a Japanese Three-Mile Island. MITI itself was criticized for insufficient supervision and announced it would review all nuclear plant safety and inspection procedures. The Nuclear Safety Commission, which has no administrative responsibility, issued its first White Paper on nuclear safety in October 1981.

Coal

Japan is currently very interested in large steam-coal imports for the electric-power industry. In 1973 coal was not given a high priority in the alternative-energy mix because of a lack of cooperation among coal producers, environmental groups, and experts knowledgeable about pollution technology. In mid-1980, the domestic Coal Mining Deliberation Council for the first time in six years evaluated Japanese options for domestic coal production and future coal imports. Domestic coal production is expensive and dangerous, and it is not expected to expand to meet new demand levels. In 1980 twenty-five domestic coal mines produced 18 million tons (of which 11.4 million was steam coal). Japan's total coal demand in 1980 stood at 91 million tons; 63 percent of the imported steam coal (4.5 million tons), and 41 percent of the coking-coal imports (26.5 million tons) came from Australia.

The Japanese government anticipates that coal imports in 1985 will provide almost 14 percent of total energy consumption. The May 1981 Ministry of International Trade and Industry coal-demand projections, reflecting a dramatic increase in steam-coal imports, are shown in table 14-2.

Conclusion

Whatever Japan's posture, or preference for separating economics and politics, the world energy crisis has forced the Japanese to realize that politics, economics, and energy are closely intertwined. On January 23, 1980, the Japanese Foreign Minister, Saburo Okita, speaking to the press about Japan's foreign policy for the 1980s, acknowledged this when he stated: "Like it or not, politics and economics are irrevocably linked in today's world." The two oil crises have demonstrated all too plainly how closely tied oil is to the political situation in the Middle East. The reason postwar Japan was able to pursue its economic diplomacy was that the broad outlines of Japanese foreign policy were stable and the Japanese economy had limited impact on the outside world. Today, however, the situation is different.

Table 14-2
Japanese Coal-Demand Projections, May 1981
(million tons)

	Year		
Type of Coal	*1980*	*1985*	*1990*
Coking coal	71	84	94
Steam coal	21	43	71-77
for electric power	10	23.5	49-55
for cement production	7	12.8	13.6
Total demand	92	127	165-171

The reappraisal of Japanese energy policy that has continued since 1973 is still unresolved. As with other nations, during a period of ample oil, the pressure to move quickly and meet energy goals is significantly reduced. Japan, like the Europeans, perceives instabilities in the Middle East as largely indigenous, and refuses to go along with U.S. views on military solutions to regional problems. Thus far the Japanese have made successful economic adjustments. Japanese realize that they are still very vulnerable to higher energy prices or possible supply disruptions. Still, they are reluctant to rely solely on multilateral solutions. Japan has gone along with but given little serious consideration to strategic coordinated multinational measures to prevent a supply crisis. It is possible that given appropriate coordination and consultation, the opportunity exists for the United States, Europe, and Japan to reduce their mutual vulnerabilities through joint efforts that would link different but complementary strategic concerns. Forging a combined consuming-nation effort will not be easy (and will require bridging differing assumptions about the nature of the energy issue), but the energy and security benefits of a combined strategy would far outweigh any unilateral actions either side might take. Coordination will be the only way to restrain future price increases. A series of agreements between Japan, the United States, and Europe would be one way to overcome the diversity and domestic energy-policy differences that have blocked effective multilateral coordination to date. As traditional alliance relations weaken, other kinds of coordination can be divisive. Japan's efforts to coordinate with Europe during the Iranian crisis were used to obtain leverage against the U.S. desire for Japan to criticize Iran.

Japanese preparedness for an energy emergency, as one would expect, is primarily concerned with enhancing Japanese economic survival. Although Japan might think it is better prepared for a major oil disruption than is the United States or Europe, its plan may prove far from adequate given the global nature of the oil market. The major policy flaw, when viewed from the perspective of the need for consuming-nation cooperation, is Japan's lack of faith in coordinated or international energy strategies. With oil demand expected to remain at the current levels through the first half of the 1980s, the Japanese are giving major attention to energy-saving options like conservation and new, alternative energies.

Notes

1. For a detailed treatment of Japan's current energy policies see: *The Politics of Japan's Energy Strategy*, Ronald A. Morse, ed. (Berkeley: University of California, 1981).

2. Japan's response to the 1973-1974 oil crisis is examined in Henry R. Nau, "Japanese-American Relations during the 1973-1974 Oil Crisis," in

Oil and the Atom, Michael Blaker, ed. (New York: Columbia University Press, 1980); and Yuan-li Wu, *Japan's Search for Oil* (Stanford: Stanford University, Hoover Institution Press, 1977).

3. Masao Sakisaka, "Japan's Long-Term Vulnerabilities," *Oil, the Arab-Israel Dispute, and the Industrial World*, J.C. Hurewitz, ed. (Boulder: Westview Press, 1976), p. 55.

4. Kazushige Hirasawa, "Japan's Tilting Neutrality," in J.C. Hurewitz, ed., *Oil*, p. 140.

5. Valerie Yorke, "Oil, the Middle East and Japan's Search for Security," *International Affairs* 57:3 (Summer 1981), pp. 428-448.

6. Walter R. Mahler, "Japan's Adjustment to the Increased Cost of Energy," *Finance and Development* (Washington, D.C.: International Monetary Fund, December 1981), pp. 26-29.

15 Energy Politics and Public Participation

Volkmar Lauber

In the 1970s, governmental energy policies came under attack in many countries. These attacks, originating not so much from opposition parties as from previously marginal groups in society at large, have taken many forms. New forms of public participation have arisen in the struggle over policy; the movements animating them have generally aimed at modifying official energy policies, particularly in the direction of reduced reliance on nuclear power and generally toward policies of low energy growth or alternative forms of energy. Governments have met the challenge, in several cases at least, by mobilizing the public on behalf of their own policies; in this they have so far met with mixed success.

Though energy policy has been influenced by the politics of public participation in many countries, the distribution of national production in those same countries is still governed by a political process in which powerful interest groups struggle for a greater share of the national product and thus cumulatively create almost irresistible pressures for economic growth (if not for inflation). This, not surprisingly, leads to a greater demand for energy.

It would be a rare coincidence if the two kinds of politics (interest-group and public-participation) were to produce identical outcomes: a conflict can be resolved in many different ways. Although the interest-group pattern of politics has so far been the more successful, a serious commitment to energy saving might challenge that pattern. This may seem an unrealistic and impossible task given the vested interests; but the challenge has already begun in the guise of public participation. What are the forms that it has taken, and what will its likely impact be?

Before the 1970s, energy policy did not create much controversy in most Western countries; nor were there strong demands for public participation in energy policymaking. In the decade of the 1960s, oil and gas expanded their hold on the energy market significantly as their prices stayed low (or decreased) throughout the decade. Increased consumption of these cheap forms of energy was connected with the goal of rapid (quantitative) growth of the economy; that goal itself was widely judged desirable and hardly controversial. Political parties and parliaments did not deal very much with energy policy. To be sure, there were some exceptions to this: strip-mining, oil-import quotas, the problems created by the decline of coal, and occasional site controversies. But hardly anyone questioned the need for abundant energy or the policy of rapid economic growth. During this period, nuclear power made its appearance on a very modest scale. Though it had once—in the 1950s—been hailed as the energy of the future (safe, reliable, and "too cheap to meter"), it had not made substantial progress because of its high cost compared with that of oil.

In the 1970s there were important changes in this pattern. To begin with, in the second half of the preceding decade ecology had already arisen as an increasingly important issue, and this was confirmed during the first half of the new decade. To many, quantitative growth of the GNP was no longer a self-evident goal of economic policy. Also, during the first years of the decade the price of oil increased: first in 1971, then most dramatically after the war and embargo of 1973-1974. Not only did the monetary cost of oil go up (something sufficiently disturbing for countries as heavily dependent on this form of energy as are most Western European countries); but also a great vulnerability also became evident.[1] This led to the acceleration of nuclear power programs in most Western European countries in order to meet the increase in energy consumption, primarily in form of electricity.[2]

In the meantime nuclear energy had become controversial. Not only was its safety criticized (reactor safety, waste disposal, fuel transportation, and so on); but also there were increasing fears as to the political structures that might become necessary as a result of a massive shift to nuclear power[3] and the police state that might be required to protect it.[4] These concerns found a large echo among many Western publics; by contrast, political parties generally tried to reduce the energy problem to the status of a nonissue even after it had become controversial. Their usual orientation was toward growing energy production and consumption as required by continued economic growth, a commitment to "hard" technologies, and a reduction of dependence on oil-exporting countries which those technologies were intended to bring about.[5]

Much of the new public participation in the field of energy policy (and particularly, though not exclusively, in the nuclear field) came from these facts: that a substantial portion of the public questioned the safety of new energy installations as well as the need for them; and that political actors (in particular political parties) rarely took up or articulated the new concerns, or did so only belatedly in response to an existing opposition movement. This has led to new forms of controversy, taking place not so much between the established political parties as between the institutions of official politics (parties, bureaucracies, industry, and organized labor) on the one hand and the public at large on the other.

Several institutional forms of this conflict shall be discussed, with the following questions in mind. Since the controversy is not only over technical issues but also over political and social ones (such as the future course of development of advanced industrial societies), the question arises as to how appropriate the various institutions are to addressing and resolving such a conflict. Do they tend to narrow the debate to technical issues, or do they allow for the inclusion of larger ones? Second, does the politics of participation produce outcomes that are likely to conflict with those produced by interest-group politics? If so, how will the conflict be resolved?

Consultation, Hearings, and Commissions of Inquiry

These are mild forms of public participation in which the state proposes to take the public into account first by giving it a right to be heard, and sometimes by also publishing the collected information and recommendations based on such hearings. The procedure that is probably the least significant in terms of public participation is the setting up of consultation procedures such as those used in France for the establishment of nuclear power plants (and generally for large industrial projects). The administrative proceedings that have to be held on such occasions include an exploration of attitudes on the part of the local population to be taken into account. A nonbinding report and recommendation are then made to the *Conseil d'Etat*, which in turn makes a nonbinding recommendation to the prime minister, who makes the decision.[6] The practical importance of these provisions is easily gauged: on all the sites where local referenda have been held, with one exception (Flamanville, France), the local population expressed its hostility to the plant, yet the government did not conceal its intention to proceed with its plans.[7] Thus a minister of industry, Michel d'Ornano, declared in 1974 that "municipalities will be consulted, but should they be opposed to the projects, then we would have to override them and impose [the nuclear power plants] in the national interest."[8] It is true though that at least in some cases local opposition brought about a delay.[9] It might be added that the administrative proceedings conducted in this context are secret, so that the official view of the attitudes of the local population remains inaccessible.[10]

Somewhat similar to consultation procedures, though potentially more significant, is the use of commissions of inquiry. The idea behind these is to gather all the relevant facts and opinions, on the assumption that this will represent a sufficient basis for an appropriate policy response. Although such commissions may help to articulate a problem, they can also obscure it.

This is well illustrated by the following two inquiries. From 1975 to 1976, an inquiry was conducted in Canada on the Mackenzie Valley Pipeline project. It must be noted that the government itself was not a party to the conflict, in which the public (particularly residents of the area) opposed

Canadian Arctic Gas Pipelines Limited. The hearings were conducted in such a way as to widen a debate that had originally been framed in narrow technical and econonic terms; it thus served "to broaden the forum of national policy debate, by explicitly revealing the nature of political and value choices implicit in the pipeline decision."[11] It was a conscious attempt to educate the public about the implications of the pipeline, and to bring their values to bear on public policy. The government report, published in 1977, became a national best-seller (and incidentally led to the selection of an alternate pipeline route). Thus the Mackenzie Valley Pipeline inquiry shows the potential of such a mechanism at its best (in terms of public participation).

But inquiries can take on exactly the opposite features, as in the Windscale inquiry, conducted in Great Britain in 1977. The construction of a nuclear reprocessing plant by British Nuclear Fuels Limited (BNFL), a state-owned corporation, was at stake. There was considerable public opposition to it, particularly as BNFL did not have a spotless record (it had failed to report a major radioactive pollution accident for some time). By the time the inquiry was established, the controversy had reached a high point. But Justice Parker, who presided over the inquiry, saw it essentially as a forum for a technical debate between proponents and opponents of nuclear power; he did not wish to deal with the merits of the governmental policy (this he said was for Parliament to decide) and thus rejected political argumentation. In the end, Justice Parker decided the issue unambiguously in favor of BNFL, and his report was approved by Parliament quickly and without much controversy. Thus the inquiry served to depoliticize and decrease public debate on Windscale, and create legitimacy for governmental policy in a somewhat dubious way (Justice Parker declared that political questions were up to Parliament, and the Parliament avoided such difficult questions by basing its position on Justice Parker's report). The opposition could not claim that it had not had a chance to state its argument, and so at least in appearance fairness had been respected.[12] The method might be psychologically gratifying but bears little resemblance to authentic public participation.

Citizen Challenges in Agencies and Courts

Judicial and quasi-judicial institutions are another form in which the controversy over energy policy has increasingly been acted out. Such challenges are common in the United States, where they have a long tradition; they are also becoming increasingly frequent in Germany, where several nuclear power projects (and temporarily even the whole nuclear program) have been halted by such proceedings.[13] The goal of opposition groups in the United

States has been quite openly to delay construction (by insisting for instance that certain impact statements had not been furnished) or to gain favorable publicity even when the outcome was hardly in doubt.[14] Apart from this political purpose, all that can be achieved by such proceedings is to force the government (or utility) to abide with existing legislation. To be sure, this can take on important dimensions, as in the case of a German administrative court that voided the construction permit for the breeder reactor at Kalkar because in its view this decision presented such a hazard to life that only legislators could take this fateful step.[15]

Courts have been viewed as a favorable forum by environmentalists who expect to find more sympathy in an institution that in their view is less responsive to economic interests than is the administrative bureaucracy.[16] On the other hand, it has been argued that the real questions being tried here are political in nature (such as safety risks versus economic growth) and should therefore be decided in a political forum, particularly since lawsuits may disguise principled objection to nuclear power as concern for other causes (such as protection of certain wildlife habitats); furthermore, litigation tends to disrupt projects at a very advanced stage.[17] No wonder then that a *Wall Street Journal* editorial asked whether an endless judicial process was "truly a rational way to address the future,"[18] or that a German minister of the economy proposed that legislation should be changed in such a way as to make further court challenges impossible.[19] Moves to restrict the right to challenge governmental decisions have resulted. In Germany there was talk of narrowing (rather than expanding, as intended earlier) the right to make court challenges; and in the United States, the Supreme Court in June 1978 interpreted the right to bring a suit in such controversies restrictively.

Home Rule: Claims to Local Self-Determination

A claim to special rights of public participation in policymaking has been advanced by local populations living close to large-scale energy projects; they experience the greatest safety risks and the greatest disruption of their lives. At times they have claimed a right to veto such projects. These arguments were put forward in Wyhl, Germany, where local farmers opposed a nuclear plant because it would destroy their way of life. They argued that clouds from the cooling towers would modify the local climate, and industrialization of the area, as connected with the plant, would obliterate their whole lifestyle. A similar argument was made by the farmers of Brokdorf.[20] In both cases, the local population asked why they should be made to carry the risk of radioactive pollution so that people in other areas could increase their energy consumption.

Nuclear power plants in Brittany have been opposed on much the same grounds by local fishermen, shellfishermen, and farmers, and a similar uproar occured earlier in the Beaujolais over a refinery that winegrowers thought represented a threat to their high-quality products (after a pollution accident from a refinery near Bordeaux, the whole crop of an area had to be destroyed).[21] The many referenda held in France concerning the siting of prospective nuclear power plants (and refineries) were all negative except for one case.[22] In some cases the municipal council or the mayor have expressed their opposition to such plants.[23] Polls in Germany have shown that in several towns a majority of the population was opposed to nuclear installations.

But neither in France nor in Germany do local communities have the power to decide whether a power plant will be located there. There is a precedent for this, however, in the United States, in New Hampshire. In 1973, Olympic Refineries wanted to build a refinery in the city of Durham. Governor Meldrin Thompson strongly favored such an installation on New Hampshire territory, and so, apparently, did the population of the state. Not so, however, the population of Durham; according to one description, the town "convulsed" at the announcement of the project.[24] In a town meeting, the local population rejected the refinery by a vote of 1,254 to 144 (the vote was without legal effect; however, it was expected that the same majority would vote against a zoning variance that the refinery needed, and that the town had the power to grant or to refuse). The governor favored a measure that would have deprived local communities of such prerogatives, and the subject came before the legislature. But two days earlier it had been debated in approximately 200 town meetings across the state. Two questions were debated by the towns: Should there be an oil refinery in the state? Should local communities have the power to make decisions about such installations? "The resulting votes showed strong support for the refinery across the state, but the importance of maintaining local control over site selection was termed of overriding importance."[25] In other words, though a refinery was thought to be desirable, local self-determination was valued even more highly. This position was endorsed by the New Hampshire legislature, and Olympic Refineries abandoned its project.

A similar system was recently discussed in Switzerland. Opponents of nuclear power had initiated a referendum that would have given local areas the right to veto nuclear power plants on their territory. After an intense campaign (in which the Swiss government, industry, and labor opposed the measure) the proposal was defeated, though only very narrowly (with 51 percent of the votes) in February 1979.[26]

There are other illustrations of what is basically the same problem. Several American states have passed resolutions or legislation prohibiting the storage of nuclear wastes on their territory. Something similar happened in Germany, where the only appropriate strata for such storage are in Lower Saxony; after strong local opposition and a public hearing, the state government rejected any such installation at the time.[27] The situation is not entirely different with regard to coal. It is conceivable that some states in the American West may not want to serve as national coal mines—something that would become almost inevitable should U.S. energy policy shift heavily toward the use of coal. What these situations have in common is that nuclear energy as well as coal are national energy resources; they can hardly be restricted, as a matter of policy, to particular areas. But the costs and benefits are distributed very unevenly; and the problem of local (or regional) self-determination arises.[28]

Can it be concluded that the power of decision over energy policy should reside with that territorial unit which corresponds to the size of the energy system, in the same way as federal power in the United States expanded as the economy became national in scope? There is considerable hesitation to put the issue in such stark terms, and to overcome local opposition purely by coercion. In some countries (even in France) local communities considered for nuclear facilities receive promises of a greatly increased tax base, additional public services and investments and so on, yet obviously those steps have not been sufficient to change the public's mind.

In most countries it would be quite possible to force a nuclear plant on a reluctant local community, at least in legal and constitutional terms; this has been done before with railroads, airports, and roads. But the political cost of such a step is quite high, and most governments hesitate to use compulsion. A notable exception is the French government, probably because of a long tradition of centralization in politics and administration and an accepted habit of resorting to riot police. In some countries much effort has been made to discredit opposition movements as the work of outsiders, subversives, Communists, and the like. Although such accusations are not always completely unfounded, they have been considerably overdone.[29]

Why is it that governments are so reluctant to impose nuclear plants on unwilling local communities? It seems quite clear that there is a likely difference between local and national *interest*, which legal and constitutional norms normally resolve in favor of the national unit. But it is not clear that there is also a conflict between local and national *will*—the ultimate source of legitimacy in democratic politics. At least in the 1970s, national energy policies based on a commitment to high rates of economic growth seemed to suffer from such a legitimacy deficit.

Several German surveys illustrate the extent to which the established authorities had lost credibility in the field of energy policy. Two such surveys showed that with regard to environmental information, spokesmen for industry (and in particular for utilities) as well as politicians were accorded the least trustworthiness; doctors and scientists, environmentalists, and spokesmen for citizen initiatives ranked highest in credibility.[30] Other surveys asked specific questions about plant siting; they also showed remarkable distrust of the administration as well as political parties and the government; citizen initiatives by contrast were viewed very favorably (see tables 15-1 and 15-2).

The construction of power plants and other energy installations is usually justified by reference to the need for continued economic growth and a rising standard of living. But even these goals no longer meet with a universal (or near-universal) consensus. A survey in 1975 showed that 61 percent of German respondents declared that they were prepared to accept a net reduction in energy consumption if necessary to protect the environment.[31] And when offered a choice between economic growth and environmental protection, an overwhelming majority of German respondents in two separate national surveys opted in favor of the environment (see tables 15-3 and 15-4); it is remarkable that the difficult economic situation did not bring more of a change in attitudes between 1974 and 1977. It should be noted that the preference for environmental protection is in fact less marked at the local level, perhaps because the benefits of economic growth were put in more immediate terms (secure jobs, wages, and standard of living). In any case, it is clear from tables 15-1 through 15-4 that local opposition to energy installations finds a good deal of support on the national level.

It becomes clear why the state and federal governments in Germany are on the whole hesitant to impose nuclear power on reluctant communities: the policy goals in the name of which nuclear power is defended are themselves no longer evident; on the contrary, they are treated with considerable skepticism. As a German contributor to the International Conference on Nuclear Power put it,

> For many people in this country, the utilization and expansion of nuclear power is a signal for a development oriented in the long run towards extensive economic growth—a development which goes hand in hand with the spoiling of our environment and with the wasting of raw materials and energy. Since the discussion on the "Limits to Growth" began, or even earlier, . . . the formula of personal prosperity for all has lost its conviction.
>
> This uneasiness (about a quantitative growth society) is still intensified by the fact that the realization of the aims cited as essential reasons for this growth policy—namely increasing private prosperity, and improved offer of public services, security of jobs and full employment—cannot be guaranteed for the foreseeable future and not even for today.[32]

Such attitudes may be widespread in a number of other Western countries.

A frequent objection to the new participatory movements that started in the late 1960s is that they are supposedly uninformed if not Luddite, parochial in outlook, and irresponsible in their attitudes.[33] In the case of

Table 15-1
National Survey on Legitimacy of Local Opposition to Energy Projects
(percentages)

Question/Response	*June 1975*	*September 1976*
When a power plant or refinery is built these days, there are often environmental organizations or citizen initiatives that make opposition. . . . With which view would you rather agree?		
Most of the time such actions delay the constructions of needed installations unnecessarily and thus endanger, and make more expensive, the supply of energy. After all, the firms are controlled by the authorities with regard to regulations on environmental protection. [Oppose local opposition].	33	27
It is good that citizens defend themselves in this way to protect themselves against environmental hazards. Otherwise the firms would not take sufficient care. [Support local opposition].	50	58

Source: *Allensbacher Jahrbuch der Demoskopie*, 1974-1976, vol. 6, p. 175; 1976-1977, p. 185 (Vienna, Munich, Zurich: Verlag Fritz Molden, 1976, 1977).

Table 15-2
Local Surveys on Legitimacy of Local Opposition to Nuclear Power Installations Where Nuclear Power Plants Were Planned.
(percentages)

Question/Response	*Ludwigshaven*	*Landkreis Emmendingen*	*Freiburg*
The government and the parties are not listening to us anyhow. For this reason we have to represent our own interests in a citizen initiative.			
Positive view of parties, government	16	14	9
Negative view of parties, government	47	63	72

Source: Battelle-Institut Frankfurt/Main, *Bürgerinitiativen im Bereich von Kernkraftwerken* (Bonn: Bundesministerium für Forschung und Technologie, 1975), p. 240.

Table 15-3
Public Opinion on Economic Growth versus Environmental Protection
(percentages)

Question	*March 1973*	*May 1977*
Today the most important thing is for economic growth to continue, even if the natural environment suffers as a result.		
Agree	14	24
Disagree	74	77

Source: Rat von Sachverständigen für Umweltfragen, *Umweltgutachten 1978* (Bonn: Bundesministerium des Innern, 1978), p. 450.

Table 15-4

Local Public Opinion on Economic Growth, Secure Jobs, and Standard of Living versus Environmental Protection *(percentages)*

Opinion	*Ludwigshaven*	*Landkreis Emmendingen*	*Freiburg*	*Weighted Average*
We need more electric energy and new energy sources. With those we can maintain economic growth and secure jobs, wages and thus our standard of living. [Prefer economic growth.]	52	35	34	41
We have to question economic growth. Now the protection of the environment has priority. What good is more and more growth if it endangers the basis for our survival? [Prefer environmental protection.]	37	54	55	48

Source: Battelle-Institut Frankfurt/Main, *Bürgerinitiativen im Bereich von Kernkraftwerken,* p. 238.

nuclear power, these claims do not seem to stand up. On the contrary, it seems that opponents are often unusually well informed even about technical details, while planners have sometimes overlooked local conditions that seem evident.[34] Even if these movements may have been parochial at one time (in the early years their only goal was often to keep power plants away), they have in recent years become increasingly national in outlook, and have been compared to political parties in this respect.[35]

A situation is of course conceivable in which the political will of the local population is in open conflict with the clear political will of the national population. (A conflict of this kind has occurred in California, where a county in 1978 voted against a nuclear power plant even though an antinuclear referendum had just lost on the state level two years earlier.)[36] But if the national authorities have to enforce a policy against the clear will of a local community (or even against the strident objection of a minority), they will feel the need to present their own policies as legitimate, to show that the nation is behind them. Precisely this has occurred in several Western European countries.

Popular Participation at the National Level

In these Western European countries the governments took steps designed to involve the public in matters of energy policy. The purpose of these steps was ambiguous. On the one hand, there was the effort to promote free and rational discussion, so that the public could form an opinion on the subject. On the other hand, some governments felt that organized discussion would strengthen support for nuclear power and sponsored it for that reason. It is clear that the programs were almost always intended to generate legitimacy for governmental policies;[37] thus the suspicion of manipulation. Programs of this kind were conducted in Sweden, Demark, Austria, and Germany.

As in most countries, nuclear power was not a prominent political issue in Sweden until the spring of 1973, when the nuclear debate "literally exploded," catching most people (including industry) quite unprepared.[38] Unlike the situation in most countries, there was a nonmarginal party (the Center Party) that took up the soon-very-popular antinuclear cause. It was in late 1973 that the government (which under the Social Democrats was strongly committed to nuclear power) decided that it was time to involve the public in this policy. Public opposition in this essentially consensus-based polity had reached such levels that political-party hegemony alone no longer seemed sufficient to support the nuclear program (a poll in early 1974 indicated that more than half of the Swedish population opposed nuclear power). Thus the government set up study circles (an institution inherited from the nineteenth century) to discuss the problem; they were sponsored by political parties, unions, churches, temperance groups, and youth organizations. About ten thousand such circles, with about a dozen members each, were set up across the country.[39] They discussed a broad range of options, including alternative energy sources and energy conservation. However, it appears that much of the emphasis was technical in nature; also, the government refused to assist nuclear opponents with the development of certain information materials such as the translation of *World Energy Strategies* by Amory Lovins.[40] The impact of the study circles is not clear: while "government officials expected that increased involvement would create more favorable public attitudes towards nuclear power . . . reports from the study groups suggested that prior commitments persisted, with some increase in uncertainty and confusion."[41]

The study groups were soon terminated, but public discussion of nuclear power persisted. Recently public attitudes in Sweden have become more favorable to nuclear power; in 1978 for the first time there was a plurality favoring it. The Three-Mile Island accident in March 1979 drastically changed the situation, but soon the previous trend returned (see table 15-5).

In Sweden a proposal had been made several years earlier to hold a referendum by the Center Party and the Communist Party (both of which were opposed to nuclear power) at a time when public opinion was more consistently antinuclear.[42] When the other parties, particularly the Social Democrats under Olof Palme, came to support this proposal after the Three-Mile Island accident, this served the purpose both of overcoming conflict within the Social Democratic Party (which had developed an antinuclear faction) and of separating the nuclear issue from the 1979 elections, to the disadvantage of the Centrists and Communists who wanted the two votes to take place simultaneously. When the referendum was held in March 1980, the electorate was presented with three choices, all of which appeared to provide for the phasing out of nuclear power; but in fact two of the alternatives made the phasing out conditional on its being compatible with the maintenance of employment and welfare. This provided a rather elastic criterion (and indeed the Swedish power industry has expressed its belief that the Swedes will gradually give nuclear power broader support), even though in the prereferendum campaign the advocates of these two proposals referred generally to a delay of approximately twenty-five years (the average life expectancy of a reactor). Thus it is possible to conclude that the Swedes indeed voted for the phasing out of nuclear power, but also that the parties that stood behind the proposals for phasing out may not feel committed to draw this conclusion (their own proposals contain only a commitment not to expand nuclear power beyond the twelve reactors already in operation, completed or under construction, but no clear commitment to phase them out within a specific time). The three alternatives presented at the referendum are summarized in table 15-6.

Table 15-5
Public Attitudes toward Nuclear Power in Sweden
(percentages)

Position on Nuclear Energy	Date of Poll									
	October 1976	*May 1977*	*September 1977*	*March 1978*	*September 1978*	*January 1979*	*Early April 1979*	*Late April 1979*	*May 1979*	*August 1979*
For	27	32	35	39	41	41	26	35	35	44
Against	57	49	46	40	37	43	53	44	46	38

Source: SIFO (Vallinby) and Swedish Secretariat for Future Studies.

Table 15-6
Proposals at the Swedish Referendum on Nuclear Power, March 1980

Alternative	*Sponsor*	*Content*	*Vote (percentage)*
1	Conservative party	Nuclear power to be phased out at the rate that is possible with regard to the need for electricity for the maintenance of employment and welfare. At most twelve reactors, no further expansion.	19
2	Social Democratic Party and Liberal Party	Beginning as above. Also advocates promotion of energy conservation, protection of weakest social groups; ban on electric heating; stronger safety and environmental standards; and a greater role for the public sector in energy.	39.3
3	Centrist and Communist Parties	Six existing reactors are to be phased out over a period of at most ten years; reactors not charged are not to be put into operation; emphasis on alternative energy forms and conservation.	38.6

Source: Per Ragnarson, "Before and After: The Swedish Referendum on Nuclear Power," *Political Life in Sweden* 5 (New York: Swedish Information Service, September 1980).

Denmark launched a study program similar to that of Sweden, stressing above all social pluralism and self-education. The first plans for a commercial nuclear power plant were announced in late 1973; within a few months this became the subject of a very intense debate throughout the country.[43] Several political parties called for a broad public debate on energy questions, and the ruling Social Democratic Party announced that a public information campaign would be undertaken. The goal of the campaign was to discuss the issue of energy primarily in political and social (rather than narrow technical) terms; to give equal participation and representation to both sides, particularly with regard to the information produced; and to involve the public as much as possible. A large number of local and regional groups obtained funds to produce relevant information materials; many of them focused on alternative energy forms rather than on nuclear energy itself. The debate involved a large number of people (about 150,000), and public awareness of the problems increased considerably. In 1976 the government decided to postpone the nuclear power program it had planned, since it did not want to act in the absence of a popular consensus (a referendum is planned for 1981). The information campaign was discontinued, but many of the groups that had organized during the campaign (and had received assistance from the government) remained in existence. The net effect of the campaign apparently has been to increase resistance to nuclear power. In early 1974, a plurality of about two to one favored nuclear energy; but by 1976 the situation had reversed itself, and in 1977 a clear plurality was opposed.[44]

In Austria an information campaign was also conducted when the nuclear program, which went back to the 1960s, became controversial in 1973 and afterward. The Austrian chancellor, strongly in favor of nuclear power, nonetheless announced that this question should not be left to experts but should be opened to public debate. In 1976 an information campaign was prepared; public debates were organized and televised nationally and efforts made to secure independent expertise by excluding from the debates all parties associated with firms involved in the Austrian nuclear program. The information campaign was completed in 1977. In early 1978, public-opinion polls indicated that public sentiment favored nuclear power; at that time the government decided to hold a referendum. In the campaign, the Social Democratic government under Chancellor Kreisky argued very strongly that nuclear power was necessary to ensure continued economic growth, employment security, and a high standard of living; the chancellor even threw his own prestige and popularity into the balance by hinting that he might resign if the outcome was negative. His position was supported by organized labor and the federation of industry, while opposed (with some ambiguity) by the Conservative Party and a considerable number of dissenters within the chancellor's own Socialist Party (particularly the youth organization). On November 5, 1978 the population voted, by a majority of less than 51 percent, to close down the Zwentendorf plant, which by that time was fully completed at a cost that exceeded $500 million.[45] The Austrian parliament (with the votes of both major parties) then terminated the whole nuclear power program, outlawed the use of nuclear power on Austrian territory, and made any change of this policy dependent on an affirmative referendum. It was not long before advocates for such a referendum made themselves heard; in the summer of 1979 such efforts were already under way (coming particularly from organized labor, which in this context is also arguing the case for industry, true to the principle of *Sozialpartnerschaft*, the Austrian version of corporatism); in November 1980 the necessary legal steps were finally taken to initiate a reversal of the 1978 referendum decision.[46]

In Germany the government sponsored an effort at public involvement at a time (late 1974) when opposition to nuclear power had not yet reached a very high level of intensity (this came with local opposition movements such as the ones in Wyhl and Brokdorf). In fact, one of the reasons for the program was to prevent the rise of such an opposition movement. The *Bürgerdialog* (dialogue with the citizens) consisted very largely of an information campaign based on materials sent out by the Federal Ministry for Research and Technology; at the same time, opinion-forming processes in large groups such as political parties, labor unions, industrialists' groups, churches, and youth groups were also encouraged. There are some doubts as to the objectivity of the materials introduced; adversaries of nuclear power characterized the whole enterprise as monologue rather than dialogue. It is clear in any case that the government throughout the time of the dialogue never wavered in its commitment to nuclear power development; that in 1977 it started to discredit and monitor citizen initiatives; and finally, that it gave little weight to its own party constituency (the rank and file of both parties of the governing coalition appeared in the main to have grave reservations about nuclear power), but encouraged expressions of support for nuclear power by groups that seemed poorly placed to make an independent and responsible judgment.[47] Most obvious of the moves to suport nuclear energy was a movement sponsored by shop stewards from energy industries, who—generously financed by such firms as MAN, Krupp, KWU, Hochtief, and Deutsche Babcock, all heavily involved in the nuclear program—forced a reluctant DGB (the German labor-union federation) leadership into endorsing nuclear power at the Dortmund conference, November 1977.[48] The actions of the German government leave the impression that their main goal with the Bürgerdialog was not the promotion of free debate (the outcome of which, in keeping with democratic principles, would be left open), but a large operation of coordination, of securing legitimacy for existing programs while only giving the illusion of public participation.

Conclusion

What are the characteristics and implications of the different forms of public participation in energy policy? At the outset, it shall be repeated that in most of the countries discussed, political parties, bureaucracies, and corporate groups such as industry and organized labor generally took a pronuclear position and thus did little to articulate the debate. Where energy policies have become controversial, the debate normally originated with citizen groups, local communities, and sometimes scientific groups. The relative absence of political parties has contributed to the legitimacy deficit that came to characterize energy policy in the mid-1970s.

Most spontaneous forms of participation took shape in opposition to specific energy-policy measures taken by governments and sometimes by business. The importance of such opposition is in the fact that not only those specific measures were questioned, but also the underlying assumptions of governmental policy were questioned. Governments tended to respond to particular challenges by invoking those goals which justified their overall policy of continued economic growth and a rising standard of living. However, this did not settle the issue as those goals themselves met with increasing skepticism. Some governments have responded to this situation by promoting mobilization of the population for their own goals, so far with uncertain success.

The question arises why most governments are so strongly committed to the policies of energy growth. A common view is that it is simple necessity, resulting from the domestic and/or international situation, that makes such a course imperative. This notion is reflected in the German term *Sachzwang,* so frequently invoked in the discussion of energy policy. The implication is that no real choice in the matter is possible; public opinion therefore becomes irrelevant (or more strictly speaking, the only task that remains is to convert public opinion to the correct point of view). Participation can therefore only serve the purpose of legitimizing predetermined policies. The implications of such a situation were noted by Jacques Ellul many years ago:

> With regard to an enterprise that involves billions and lasts for years . . . such as French oil policy in the Sahara or electrification in the Soviet Union, public opinion can play no role whatsoever.
>
> Ergo: even in a democracy a government that is honest, serious, and respects the voter *cannot follow* public opinion. But it *cannot* escape it either. . . . Only one solution is possible: as the government cannot follow public opinion, public opinion must follow the government. One must convince this present, ponderous, impassioned mass that the government's decisions are legitimate. . . . This is the great role that propaganda must perform.[49]

Propaganda fulfills this function not by brainwashing but by giving the public at least the impression of participation, "the immense satisfaction of having been consulted, of having been given a chance to debate, of having . . . their opinions solicited and weighed"—one might add: of having had an opportunity to make a protest, to stage a sit-in or even to confront the riot police or national guard.[50] All these may in fact serve the function of rituals; they discharge emotion but all too easily remain ineffective when it comes to policymaking.

If one assumes that no real choice is possible, then such an analysis is indeed compelling. If one concludes, however, that a margin of choice is available, one may wonder at the gap that seems to separate the hard commitments of governmental choices from the preferences and ambiguities of public opinion. Whatever the necessities dictated by intenational competition between states, one might surmise that at least with regard to the domestic situation there may be some leeway for choices (and significantly, it seems to be the smaller nations that appear to be more open to such reflection). It is certainly true that energy policy is never made in a void, that on the contrary it is very largely a by-product of the economic conditions and policies prevailing in a given country. These in turn are shaped decisively by powerful interest groups that strive for a greater share of the GNP. The most appropriate response of the system to this situation has been a politics of maximum feasible economic growth. It is unclear what else the system has to offer if economic growth should cease.

It is only normal that a political process shaped by interest groups governing the economic sphere will come into conflict with a process based on public participation (as long as participation is not manipulated), a conflict which has been taking shape, to a greater or lesser extent, in the field of energy. Parallel outcomes of the two processes would be highly coincidental. If economies of energy, and of other resources as well, should become central policy goals, then the greater obstacle in the way of such goals and policies in the Western countries may not be the stubborness of their citizens but the structures of their political economies. It is these economic structures that will have to be confronted; undoubtedly an immense, perhaps an impossible task. Yet the challenge to those structures may already be on the way, in the form of the politics of public participation.

Notes

1. Guy de Carmoy, *Energy for Europe* (Washington, D.C.: American Enterprise Institute, 1976); and Wilfrid L. Kohl's chapter on the European Communties in this book.

2. The most extreme case was France, where plans in 1977 were to increase the contribution of nuclear energy from supplying 1.8 percent of French energy needs in 1973 to 25 percent in 1985 (covering 70 percent of all electric production at that date). The average for the United States and Western Europe projected then was around 30 percent of electric production. *Rapport Schloesing* (Paris: Assemblée Nationale, November 15, 1977), p. 18.

3. Alvin Weinberg discusses an "electric utopia" of several thousand breeders in "Reflections on the Energy Wars," *American Scientist* (March-April 1978), p. 157. For even higher projections see Mihajlo Mesarovic and Eduard Pestel, *Mankind at the Turning Point* (New York: Dutton, 1974), pp. 131-135.

4. Harry J. Otway, Dagmar Maurer, and Kerry Thomas, "Nuclear Power: The Question of Public Acceptance," *Futures* 10:2 (April 1978), pp. 109-118.

5. Leon Lindberg, *The Energy Syndrome* (Lexington, Mass.: LexingtonBooks, 1977), p. 338.

6. Jean-Philippe Colson, *Le nucléaire sans les Français* (Paris: Maspéro, 1977), pp. 109-112.

7. Pierre Samuel, *Le nucléaire en questions* (Paris: Entente, 1975), p. 100; and the Electricité de France memo in Colson, *Le nucléaire,* pp. 183-184.

8. Quoted in *Combat Nature* 27 (February 1977), p. 45.

9. As for instance in Brittany: Martine Chaudron and Yves Le Pape, *Le mouvement écologique dans la lutte anti-nucléaire* (Paris/Grenoble: ADISH/IREP, 1977), pp. 3-39.

10. The French Friends of the Earth tried to obtain such documents under the new law of July 1978 (a kind of freedom of information act); so far unsuccessfully. *Le Nouvel Observateur*, March 26, 1979, p. 27.

11. *Technology on Trial* (Paris: OECD, 1979), pp. 68-77, and references there, in particular T.R. Berger, *Northern Frontier-Northern Homeland: Report of the Mackenzie Valley Pipeline Inquiry* (Ottawa: Information Canada, 1977); and D.J. Gamble, "The Berger Inquiry: An Impact Assessment Process," *Science* 199 (March 3, 1978), pp. 946-952.

12. *Technology on Trial,* pp. 63-68.

13. "Nuclear Wastes Stymie West Germans", *Science* 195 (March 11, 1977), p. 962. As a result of that court decision, the government promotes the idea of an interim solution to the storage problem that should allow, it argues, continuation of the program.

14. Irving Like, "Multi-Media Confrontation: The Environmentalists' Strategy for a 'No-Win' Agency Proceeding," *Ecology Law Quarterly* 1 (1971), pp. 495-518.

15. *Der Spiegel,* August 22, 1977, p. 73 and October 3, 1977, pp. 124-130.

16. Roger C. Cranton and Barry B. Boyer, "Citizen Suits in the Environmental Field: Peril or Promise?" *Ecology Law Quarterly* 2:3 (Summer 1972), p. 410.

17. Ibid., p. 412.

18. *The Wall Street Journal,* February 22, 1977, p. 33.

19. *Der Spiegel,* March 21, 1977, p. 33.

20. Volkmar Hartje and Meinolf Dierkes, "Impact Assessment and Participation: Case Studies on Nuclear Power Plant Siting in West Germany," (Berlin: Science Center/Institut für Umwelt u. Gesellschaft, 1976); and Battelle-Institute Frankfurt/Main, *Bürgerinitiativen im Bereich von Kernkraftwerken* (Bonn: Bundesministerium für Forschung und Technologie, 1975); *Bürgerinitiativen/Bürgerprotest: eine Neue Vierte Gewalt* (Berlin: Kursbuch Verlag, 1977).

21. Chaudron and Le Pape, *Lemouvement écologique,* p. 17; Christian Garnier-Expert, *L'Environment démystifié* (Paris: Mercure de France, 1973), pp. 71-73. In French polls 1972-1975, refineries were even less popular as potential neighbors than nuclear power plants; see Electricité de France *Le public à l'heure nucléaire* (Paris, 1975).

22. See note 10. A petrochemical complex in Château d'Oléron was also the subject of a referendum: 66 in favor, 1052 opposed. *Le Monde,* June 16, 1976, p. 37.

23. *Combat Nature,* February 1977, pp. 44-49, gives a list of such municipalities.

24. J. Douglas Peters, "Durham, New Hampshire: A Victory for Home Rule?" *Ecology Law Quarterly* 5, pp. 53-67.

25. Ibid., p. 59.

26. *The Wall Street Journal,* February 20, 1979, p. 12. According to the report, the right to decide would fall to the people living within a twenty-mile radius.

27. Luther J. Carter, "Nuclear Wastes: Popular Antipathy Narrows Search for Disposal Sites," *Science* 197 (September 27, 1977), pp. 1265-1266 (negative reactions came from Michigan, Louisiana, South Carolina, Georgia, Vermont, and South Dakota); *Der Spiegel,* June 4, 1979, p. 27.

28. Local opposition may mean opposition to a particular installation, or to a particular energy form per se.

29. William Sweet, "The Opposition to Nuclear Power in Europe," *Bulletin of the Atomic Scientists* (December 1977), p. 43.

30. Rat von Sachverständigen für Umweltfragen, *Umweltgutachten 1978* (Bonn: Bundesministerium des Innern, 1978); and *Resumée einiger Wichtiger Ergebnisse der Umweltschutzuntersuchung* (Bonn-Bad Godesberg: Institut für Angewandte Sozialforschung, 1977), p. 12.

31. EMNID (Bielefeld: March 1975).

32. K. Lang and M. Popp, "The Public Discussion about the Peaceful Utilization of Nuclear Energy in the Federal Republic of Germany and the Information and Discussion Campaign of the Federal Government," paper given at International Conference on Nuclear Power and Its Fuel Cycle, Salzburg, Austria, May 2-13, 1977. (Conference proceedings published by IAEA, Vienna, Austria); pp. 9 and 12.

33. Samuel P. Huntington, "Postindustrial Politics: How Benign Will It Be?" *Comparative Politics* 6:2 (1974), pp. 163-191.

34. Lang and Popp, "Public Discussion in Germany," p. 12; and Peter Cornelius Mayer-Tasch, *Die Bürgerinitiativbewegung* (Reinbek: Rowohlt, 1976), p. 81.

35. Rat von Sachverständigen für Umweltfragen, *Umweltgutachten 1978,* p. 461; Battelle-Institut, *Bürgerinitiativen,* p. 22; Hartje and Dierkes, "Impact Assessment," p. 21.

36. Lettie McSpadden Wenner and Manfred Wenner, "Nuclear Policy and Public Participation," *American Behavioral Scientist* 22:2 (November-December 1978), p. 307.

37. In three of the four countries, strong Social Democratic leaders (Willy Brandt, Olof Palme, and Bruno Kreisky) felt that one of the most important goals of the Social Democrats today was to defend industrial society against "irrational attacks." Dorothy Nelkin and Michael Pollak, "The Politics of Participation and the Nuclear Debate in Sweden, the Netherlands, and Austria," *Public Policy* 25:3 (Summer 1977), pp. 353-354.

38. Ibid., pp. 333-357; and Lennart Daléus, "A Moratorium in Name Only," *Bulletin of the Atomic Scientists* (October 1975), pp. 27-33.

39. Nelkin and Pollak, "Politics of Participation," p. 343. Dorothy Nelkin, *Technical Decisions and Democracy* (Beverly Hills: Sage, 1977); *Technology on Trial,* pp. 27-31.

40. Daléus, "Moratorium," p. 30.

41. Nelkin and Pollak, "Politics of Participation," p. 344.

42. Mans Lönroth, *The Politics of the Back End of the Nuclear Fuel Cycle in Sweden* (Stockholm: Secretariat for Future Studies, 1979), p. 1.

43. *Technology on Trial,* pp. 36-40.

44. Observa Institute (Copenhagen). See *Technology on Trial,* p. 38.

45. *Technology on Trial,* pp. 34-35; Nelkin and Pollak, "Politics of Participation," pp. 349-353; H. Hirsch and H. Nowotny, "Information and Opposition in Austrian Nuclear Energy Policy," *Minerva* 15 (1977), pp. 314-334; *Der Spiegel,* November 12, 1978, pp. 118-121 and 133-139; and several articles in the *Österreichische Zeitschrift für Politikwissenschaft* 9:1 (1980).

46. On the discussion surrounding the initiatives to reverse the result of the first referendum and the questions this raises about the credibility of

democratic institutions, see Norbert Leser, "Atomare Sprengkraft und Stabilität der Demokratie," *Die Zukunft* (September 1980), pp. 10-12; and by the same author, "Ein Umbruch des Parteiensystems Steht Bevor," *Die Presse,* November 22-23, 1980, pp. 5-6.

47. *Technology on Trial,* pp. 31-34; Lang and Popp, "Public Discussion in West Germany"; Bundesministerium für Forschung and Technologie, *Kernenergie: Eine Burgerinformation* (Bonn: Bundesministerium für Forschung und Technologie, 1976); *Die Zeit*, August 12, 1977, p. 3.

48. Dorothy Nelkin and Michael Pollak, "Political Parties and the Nuclear Energy Debate in France and Germany," *Comparative Politics* vol. 12, no. 21 (January 1980); *Der Spiegel,* November 21, 1977, pp. 44-46 and December 18, 1978, pp. 52-60.

49. Jacques Ellul, *Propaganda* (New York: Random House, 1965), pp. 125-127. Similarly, with specific reference to the Swedish nuclear energy situation, Jacques Ellul in *Le Système Technicien* (Paris: Calmann-Lévy, 1977), p. 331.

50. Ellul, *Propaganda,* p. 131. See also Carole Pateman, *Participation and Democratic Theory* (Cambridge, England: Cambridge University Press, 1970), pp. 68-69.

About the Contributors

Fadhil J. Al-Chalabi has been the deputy secretary general of OPEC since 1978. He previously held high positions in the Ministry of Oil of Iraq and was assistant secretary general of OAPEC. Presently the editor in chief of the *OPEC Review*, Dr. Al-Chalabi is also the author of many articles on energy and oil economics. His book, *OPEC and the International Oil Industry: A Changing Structure*, was published in 1980.

Paul S. Basile is currently director of corporate planning for the International Energy Development Corporation (IEDC), a Swiss-based, private, energy company active in energy development in the Third World. Before joining IEDC, Mr. Basile was a policy analyst with the staff of the Subcommittee on Energy and Power of the U.S. House of Representatives. As assistant leader of the Energy Systems Program at the International Institute for Applied Systems Analysis (IIASA) in Austria and program officer with the Workshop on Alternative Energy (WAES), he has been actively involved in global energy studies.

Umberto Columbo is president of the Italian nuclear authority, the Comitato Nazionale per l'Energia Nucleare (CNEN). An engineer, he was previously in charge of research and development at the Montedison Company in Italy. He has also taught at The Johns Hopkins University Bologna Center and during 1972-1975 served as chairman of the Committee for Scientific and Technological Policy of the OECD. A member of the Trilateral Commission, he was also president of the European Industrial Research Management Association from 1977 to 1979. The coauthor of *Beyond the Age of Waste* (1978) and *Reducing Malnutrition in Developing Countries* (Trilateral Commission Paper, 1978), he also contributed the section on Italy to *World Coal Study* (1980).

Guy de Carmoy, a professor at the European Institute of Business Administration (INSEAD) in Fontainebleau, has also taught at the Institut d'Etudes Politiques in Paris. He has served as a senior civil servant with the French Treasury and with the Organization for European Economic Cooperation. His publications include *The Foreign Policies of France, 1944-1968* (1970); *Le Dossier Européen de l'Energie* (1971); and *Energy for Europe* (1977).

Georges Delcoigne is director of the Division of Public Information at the International Atomic Energy Agency (IAEA) in Vienna. He has also served in the Belgian diplomatic service.

Neils de Terra was until recently assistant director of the Office of Energy Research, Development and Technology Applications, at the International Energy Agency (IEA) in Paris. He previously worked in the U.S. Departments of Transportation and Treasury.

John F. Helliwell has been professor of economics at the University of British Columbia in Vancouver since 1967. He previously taught at Oxford University. Dr. Helliwell has worked on macroeconomic and energy models of the Canadian economy and the economies of several other countries, and he has published widely on energy and other economic questions.

Volkmar Lauber, until recently assistant professor of government at West Virginia Wesleyan College, has just been appointed professor of political science at the University of Salzburg. He has also taught at The Johns Hopkins University Bologna Center. The author of articles on political economy, ecology, and French politics, he is working on a book on the making of economic policy in France.

N.J.D. Lucas is lecturer in energy policy at the Imperial College of Science and Technology, London. The author of *Energy in the European Communities* and *French Energy, Politics, Planning and Policy*, he has written articles on energy planning, energy policies in industrialized and developing countries, and energy use in industry. He is also a member of the Council of the Institute of Energy.

William F. Martin, until recently special assistant to the executive director of the International Energy Agency, is now special assistant to the undersecretary for economic affairs, U.S. Department of State. Earlier he was associated with the Workshop on Alternative Energy Strategies (WAES) at Massachusetts Institute of Technology. He has published a monograph and several articles on international energy issues.

Ronald A. Morse is director of the East Asia Program at the Woodrow Wilson International Center for Scholars in Washington, D.C. He has a doctorate in Japanese studies from Princeton University. Before joining the Wilson Center in 1981, he held several positions in the U.S. government, the most recent being with the Department of Energy. He is editor of *The Politics of Japan's Energy Strategy* (October 1981).

Dieter Schmitt, associate professor of energy economics at the University of Cologne, Germany, since 1974, is also head of the Institute of Energy Economics at that university. He is an active consultant to government and industry and has published widely on energy economics and energy policy.

Ian M. Torrens, who until recently headed the oil-industry division of the International Energy Agency, now directs work on environmental problems connected with energy and resource development at the OECD, Paris.

About the Editor

Wilfrid L. Kohl is associate professor of international relations and director of the International Energy Program at The Johns Hopkins University School of Advanced International Studies (SAIS) in Washington, D.C. He previously served as director of The Johns Hopkins University European Center in Bologna. From 1972-1976 he was associate professor of political science and associate director of the Institute on Western Europe at Columbia University. He has also served on the staffs of the National Security Council and The Ford Foundation and was an International Affairs Fellow of the Council on Foreign Relations and the Woodrow Wilson International Center for Scholars. Recent publications include: *Western Energy Cooperation* (monograph, Royal Institute of International Affairs, London); and, as editor with Giorgio Basevi, *West Germany: A European and Global Power* (Lexington Books, 1980). He also edited *Economic Foreign Policies of Industrial States* (Lexington Books, 1977). The author of *French Nuclear Diplomacy* (1971), he has published articles on the U.S. foreign-policy process, U.S.-European relations, European security, and international energy questions.